KEY NOTES ON HORTICULTURE AND AGRICULTURE EXTENSION

For Ready Reference to the

STUDENTS, TEACHERS, RESEARCHERS & ASPIRANTS OF COMPETITIVE EXAMINATIONS

THE EDITORS

Dr. U.D. Chavan obtained his M.Sc. (Agri. in Biochemistry) degree from Mahatma Phule Krishi Vidyapeeth, Rahuri. He received his Ph.D. degree in Food Science from Memorial University of Newfoundland St. John's Canada in 1999. He has done International Training on "Global Nutrition 2002" at Uppsala University Uppasala, Sweden in 2002. Dr. Chavan worked as Senior Research Assistant in the Department of Biochemistry & Food Science and Technology at MPKV Rahuri from 1988 to 2000. During his Ph.D., he worked as Technician/Research Associate at Atlantic Cool Climate Crop Research Center and Agriculture and Agri-Food Canada. He received D.Sc. degree in 2006 from USA.

Dr. Chavan is presently working as a Senior Cereal Food Technologist in the Department of Food Science & Technology at Mahatma Phule Krishi Vidyapeeth, Rahuri.

Dr. J.V. Patil obtained his M.Sc. (Agri.) from, MPKV, Rahuri. He completed his course work for Ph.D. at CCSHAU, Hisar and research at MPKV, Rahuri in 1992. He rendered his research and teaching services at MPKV Rahuri as Geneticist, Associate Professor, Plant Breeder and Professor of Genetics & Plant Breeding and Head, Genetics and Plant Breeding Department, MPKV, Rahuri. He also delivered many administrative responsibilities in the University. Dr. Patil joined as the Director, Directorate of Sorghum Research, Hyderabad in August 2010.

THE CONTRIBUTORS

Dr. K.G. Shinde is an Assistant Professor in the Department of Horticulture at Mahatma Phule Krishi Vidyapeeth, Rahuri.

Dr. A.M. Chavai is working in the Department of Agriculture Extension at Mahatma Phule Krishi Vidyapeeth, Rahuri.

Dr. S.B. Shinde is a Professor in the Department of Agriculture Extension at Mahatma Phule Krishi Vidyapeeth, Rahuri.

KEY NOTES ON HORTICULTURE AND AGRICULTURE EXTENSION

For Ready Reference to the

STUDENTS, TEACHERS, RESEARCHERS & ASPIRANTS OF COMPETITIVE EXAMINATIONS

Editors

U.D. CHAVAN
&
J.V. PATIL

Contributors

K.G. SHINDE
A.M. CHAVAI
S.B. SHINDE

2015

Daya Publishing House®
A Division of
Astral International (P) Ltd
New Delhi 110 002

Published by   : **Daya Publishing House®**
                          A Division of
                          Astral International Pvt. Ltd.
                          – ISO 9001:2008 Certified Company –
                          4760-61/23, Ansari Road, Darya Ganj
                          New Delhi-110 002
                          Ph. 011-43549197, 23278134
                          E-mail: info@astralint.com
                          Website: www.astralint.com

Laser Typesetting   : **Twinkle Graphics, Delhi**

Printed at   : **Thomson Press India Limited**

PRINTED IN INDIA

PREFACE

India is an agricultural country. The Indian economy is basically agarian. Inspite of economic and industrialization, agriculture is the backbone of the Indian economy. As Mahatma Gandhi said "India's lives in villages and agriculture is the soul of Indian economy". Agriculture is a vast subject and encompasses at least 20 major and minor subjects in it. New developments have lead to entirely a new face of agriculture. Study of agriculture has always been intrigued with a mosaic of interwove concepts, subjects, facts and figures. There are number of books and large literature on Horticulture and Agriculture Extension but the Key Notes type of book have not been compiled in a readable manner.

The present book *"Key Notes on Horticulture and Agriculture Extension"* has been designed to fulfill this long felt need of students, teachers, researchers and aspirants of competitive examinations. It is designed in such a way that give rapid, easy access to the core materials in a short format which facilitates easily learning and rapid revision. The book carries fundamentals of Horticulture and Agriculture Extension. The book is divided in two parts. The Part A of the book is Key Notes on Horticulture and the Part B of the book is on Key Notes on Agriculture Extension. The most recent information is provided along with a detailed list of references for further reading.

Hope this book would be highly useful for graduate and post-graduate students of agriculture, teachers and researchers. This book will also useful for the aspirants of various competitive examinations such as Agricultural Research Service (ARS), ICAR- National Eligibility Test (NET), State Eligibility Test (SET), Junior Research Fellowship (JRF), Senior Research Fellowship (SRF), Civil Services, Allied Agricultural Examinations and Extension Workers for reference and easy answers of many complicated questions. Thus it is expected that this book will adequately meet the need of wider circle of students and readers for preparing their professional career.

We acknowledge the references that are used in this manuscript. Authors are also thankful to all scientists and friends who have helped directly or indirectly while preparing this manuscript. The editors of grateful to all the contributors for their cooperation, support and timely submission of their manuscripts for bringing out this publication. We would have like to acknowledge the patience and support of our families whilst we have spent many hours with drafts of manuscripts rather than with them. Lastly, our sincere thanks to publisher Astral International Pvt. Ltd., New Delhi who provides an opportunity to publish this book.

To all readers we extend an invitation to report that no doubts have escaped our attention and to offer suggestion for improvements that can be incorporated in future editions.

U.D.Chavan and J.V. Patil

Editors

CONTENTS

KEY NOTES
ON
HORTICULTURE

1

TERMINOLOGY

Term	Terminology
Acclimatization	Adjustment in tolerance shown by a species in the course of several generations in a changed environment, adapting to a new set of conditions.
Aggregate fruit	Made up of two or more carpel from a single flower and the stem axis, e.g., Blackberry, strawberry.
Allogamy (cross pollination)	Pollen grains from flower of one cultivar/species pollinate the flower of other cultivar or species.
Alternate bearing	Biennial bearing; Bearing heavy and lean crop in alternate years, e.g., mango. Good cultural practices can change 'off ' years into 'on' years.
Annual	A plant that completes its life cycle within a growing season, e.g., poached egg flower (*Limnanthes douglasii*)
Arid	Potential water losses by evaporation and transpiration more than precipitation.
Aril	An additional covering formed on certain seeds by expansion of the stalk of the ovule after fertilization.
Asexual propagation	It consists of (1) Production of bulbs and their related structures.(2) Growing the plants on their own roots as in case of cuttage and layerage (3) Growing plants on the specialized root-stocks as budding and grafting.
Auto gamy (self pollination)	A highly self pollinated condition where cross pollination is less than 5 per cent.
Biennial	A plant with a life cycle that spans two growing seasons e.g., Foxglove (*Digitalis purpurea*).
Bonsai	Japanese art of growing miniature trees and shrubs by extreme dwarfing.

Term	Terminology
Bowers	It is small round place of relaxation in garden. It is made-up of wood, having sites open and roof is covered with good climbers.
Budding	Budding is the vegetative method of plant propagation. It is an art of incretion of a single mature bud into the stem of the root stock in such way that the union takes place and the combination continuous to grow.
Bulk pruning	Removal of large branches as contrasted to removal of small branches.
Candy	Fruit or vegetable product, a preserve which has been heavily impregnated with sugar.
Caprification	The process of pollination of the edible fig with pollen from the Caprifig by the Blastophaga psens, wasp.
Caprifig	The wild or uncultivated fig used to pollinate the domestic edible fig.
Classification	Any system involves the grouping of organisms or objects using characteristics common to members within the group.
Climacteric fruit	Fruits in which the respiration rate is minimum at maturity and remains rather constant even after harvest which gradually increases at the beginning of ripening followed by a sharp rise to a peak (climacteric peak) and then slowly decline (post climacteric stage).
Climacteric peak	The maximum point of respiration rate of mature fruit.
Clonal Selection	A breeding method of asexually propagated plants based on the selection of superior clones from a wide range of clones on visual observation at morphological traits.
Collective fruit	Fruit formed form a complete inflorescence or from several flowers as of mulberry, pineapple, Ficus.
Colne	A group of plants originated as asexual progeny of a single mother plants.

Term	Terminology
Companion cropping	It is the planting of two or more than two crops on same piece of land in same season which have similar cultural requirements, e.g., Cole crops.
Companion crops	Two non-competitive crops grown in the same area at the same time, one is of short term and the other of long duration.
Cracking	A disorder where fruit surface cracks mainly due to heavy irrigation or rain after long dry spell. This may occur due to varietal characters and micro nutrient deficiencies; generally found in tomato, litchi cherry, apple, pomegranate, etc.
Crop potation	Systematic arrangement of crop on same piece of land in proper sequence is called crop rotation.
Crotch	The angle made by the attachment of a branch to the trunk or another branch.
Crown grafting	A method of grafting in which a mature dormant wood of the previous season growth is grafted on the stock in the crown region, i.e., the transition region between the stem and root.
Cultivar	An assemblage of cultivated plants which is clearly distinguished by any character and which when reproduced, sexually or asexually, retains the raising crops.
Cuttage or cutting	Cutting is one of the vegetative method of plant propagation, where in branch is first detached from mother plant and then induced to produce roots.
Day-neutral plant	Plants in which flowering is not influenced by day-length.
Deblossoming	Removal of flowers from the plant to reduce the crop load, done either by with holding water or by spraying of chemicals or mechanical means.
Deciduous	A plant that sheds all its leaves at once, e.g., mock orange (*Philadephus delavayi*).
Degree day	A heat time unit representing one degree of temperature above a given mean daily temperature on a given day.

Term	Terminology
Degreening	The process of decomposing the green pigment in fruits by applying ethylene (1000 – 2000 ppm) or similar metabolic inducers to give a fruit to its characteristic colour as preferred by consumers, generally followed in citrus fruits but also practiced in banana and mango.
Dehorning	A severe heading back of major branches of a tree.
Dichogamy	The condition in which the male and female flowers of a plant do not mature at the same time, i.e., non-synchronization of pollen dehiscence and stigma receptivity.
Dicliny	The condition in which male and female organs are separate, in different flowers.
Differentiation	The process of changes in composition, structure and functions of cells and tissues during growth.
Dioeciously	Staminate and pistillate flowers on different plants.
Drupe	A fruit derived entirely from on ovary, one seeded, with an excocarp, fleshy mesocarp, and stony endocarp (as peach, cherry and plum.)
Edges	Any material use to divide the beds and borders from roads, paths, drives which demarcates and determines the area allotted for particular purpose, e.g., bricks and live plants like portulaca.
Emasculation	Removal of immature anthers from hermaphrodite flowers.
Ephemeral	A Plant that has several life cycles in a growing season and can increase in numbers rapidly, e.g., Groundsel.
Evergreen	A plant retaining leaves in all seasons, e.g., Aucuba (*Aucuba japonica*).
Fine pruning	Removal of small branches or twigs over the entire plant as contrasted to large branches.
Flower	Structure for sexual reproduction in plants Also the attractive colourful bloom for aesthetic purpose.

Term	Terminology
Formal garden	It is entirely formals nature and half design exactly matches to the remaining half. Bilateral symmetrical is layout arrangement and colour combination is the main characteristic, e.g., Brindawarn garden, Marshal garden in Delhi.
Fruit	The edible product of a woody or perennial plant which in its development is closely associated with a flower (botanically, a fruit is a ripened ovary).
Garden	A garden may be defined as a place for growing plants, exhibits various forms of plant life, which are consciously directed for ornamental or practical use or both.
Gene sanctuary	An area within the centre of diversity of a partialar plant species which is protected from human interference.
Girdling	Process of removing a strip of bark from branches or stem which blocks the downward translocation of carbohydrates, hormones beyond that constriction and rather accumulate above it.
Grafting or Graftage	It can be defined as an art of insertion of a scion into the stem of the rootstock in such a way that union takes place and the combination continues to grow.
Guttation	The exudation of water from the ends of the vascular system at the margins of leaves under humid condition.
Hardening	Subjecting plants to adverse conditions to hasten tissue maturation for increasing hardiness.
Heavy pruning	Removal of a large amount of wood from the plants.
Hedges	It is the continuous line of shrubby plant grown on border of the garden and kept neat tidy end impenetrable by regular pruning. It is common garden feature.
Herbaceous perennial	A perennial that loses its stems and foliage at the end of the growing season, e.g., Michaelnas daisy Aster sp. and hop (*Humuls lupulus*).

Term	*Terminology*
Hexagonal system	A system of planning in which the tress are planted at the corners of an equilateral triangle and thus six trees form a hexagon with the seventh tree at the centre.
High density orchard	An orchard where fruit trees are planted at closer spacing to get higher production per unit area without impairing the fruit quality.
Horticultural maturity	The stage of development of a plant or plant part which possesses the necessary prerequisites for utilization by consumer for a particular purpose and it depends on the particular crop.
Horticulture	It is Latin word; Hortus – Garden; Culture – cultivation. Thus Horticulture is cultivation of garden crops. It has four branches.
Indexing	Detection of virus infection of a plant by grafting or budding it on to an indicator plant known to be highly susceptible to that virus infection.
Indicator plant	A plant, which reflects specific growing condition, like deficiency of nutrient, disease invasion, etc. by its growth characteristics.
Informal garden	1t this free style of gardening and most city gardens are of this type. It is mixture of both formal and natural style. All type of diversified garden features is accommodated.
Intercropping	When two or more crops are taken on same piece of land and in same season, it is known as intercropping.
Interior landscaping	Is the provision of semi permanent plant arrangements inside conservatories, offices and many public buildings and involves the skills of careful plant selection and maintenance.
Landscape gardening	It is beautiful combination of natural areas on the earth. This is natural style and there is limitation of nature. There is no geometric symmetry in layout and arrangement but no natural situation is disturbed.

Term	*Terminology*
Lawn	Lawn is open space with hardy perennial herbaceous grass. It is important feature of garden. It is also called as heart of the garden.
Layerage	It is the vegetative method of plant propagation where in vegetative parts like stem (branches) of the plant is first induced or forced to produce roots while still attached to the mother plant.
Leader	The most preeminent and upright branch through the centre of the tree which tends to dominate all others.
Mother plant	A seed bearer, a plant from which vegetative portions have been selected for propagation.
Nomenclature	Which deals with naming?
Olericulture	It means study of vegetable crops
Ornamental gardening	It has two parts (a) Floriculture : It means study of flowers (b) Land scape gardening : It consists of planting, arrangement of homes, farm sheds, public areas, business establishment, play gardens, etc.
Papain	The latex exudates from the unripe papaya fruit having proteolytic activity.
Perennial	A plant living through several growing seasons.
Pergola	Pergola is series of arches made up of strong support iron pole are generally fixed firmly into ground and connected with ironnating. It serves as support to beautiful climbers and is situated over paths or in entrance.
Photosynthesis	It is the process in chloroplasts of the leaf and stem cells by which a green plant manufactures food in the form of high energy carbohydrates such as sugars and starch using light as energy.
Plant propagation	It is defined as the controlled reproduction of plants to perpetuate or to multiply the selected individuals or group of individual plants, which have specific value and commercial importance.

Term	Terminology
Plantation crops	The crops which have industrial value and require processing before consumption are known as plantation crops. The plantation crops requires more manual power and needs care and maintenance throughout the year, e.g., coconut coffee, areca nut, tea and spices.
Pomology	It means study of fruit crops.
Preservation of fruits and vegetables (PHT)	It is industrial part of horticulture, in which the principles and methods of preservation of different products of fruits and vege-tables are studied.
Protected cropping	Enables plant material to be supplied outside its normal availability, e.g., chrysanthemums all the year round, tomatoes to a high specification over an extended season, cucumbers from an area where the climate is not otherwise suitable. Plant propagation, providing seedlings and cuttings, serves outdoor growing as well as the green house industry. Protected culture, mainly using low or walk-in polythene tunnels, is increasingly important in he production of vegetables, salads, bedding plants and flowers.
Public park	This is informal garden but in additional animals, birds, reptiles and sea animal and inter Florence such is horse ride, elephant ride, boating, merry go round facilities are provided.
Rejuvenation	Stimulation of new growth on old plants usually accomplished by pruning.
Respiration	Is the process by which the food matter produced by photosynthesis is converted into energy usable for growth of the plant or is the process by which sugars and related substances are broken down to yield energy, he end products being carbon dioxide and water.
Riley cropping	Planting of subsequent crop when previous crop is under harvesting condition on same piece of land is called as Riley cropping.
Rockery	The term rockery is associated with large shade tree under mound of earth heaped in which boulders are embedded.

Term	Terminology
Scaffold branch	The main branches arising directly from the trunk of the tree.
Scion	It is the short piece of detached shoot containing two or more dormant buds, which when united with the stock, comprises the upper portion of the graft and from which will grow the stem and branches of the new grafted plant.
Senescence	The last stage of development when anabolic biochemical processes give way to catabolic processes leading to death of the tissue.
Sexual propagation	Consist of production of seeds, the resultant of fusion of male and female gametes and developing new seedling from such seeds.
Shoots	New growth which bears leaves.
Shrub	A woody perennial plant having side branches emerging from near ground level. Up to 5 mt tall, e.g., Lilac (*Syringa vulgaris*).
Species	Is the basic unit of classification and is defined as a group of individuals with the greatest mutual resemblance, which are able to breed amongst themselves.
Spir	A shoot or twig of limited growth.
Stock (Rootstock)	It is the lower portion of the graft which develops into the root system of the grafted plant.
Succession cropping	The systematic arrangement of crop on same piece of land in only one year is called as succession cropping.
Systematies	Which identifies the groups to be used in the classification?
Taxonomy	Which deals with the principles on which a classification is based?
Topiary work	This is an art of shaping hedges and certain plants into different shapes such as animals, birds and different geometrical shapes.
Transpiration	Is the loss of water vapour from the leaves of the plant?

Term	Terminology
Tree	A large woody perennial unbranched for some distance above ground. Usually more than 5 mt, e.g., Horse chestnut (*Aesculus hippocastanum*).
Trunk	The main axis of the plant from ground level to the point of branching.
Turf culture	Which includes decorative lawns and sports surfaces for football, cricket, golf, etc. Landscaping, garden construction and maintenance which involve the skills of construction together with the development of planted areas (soft landscaping). Closely associated with this sector is ground maintenance, the maintenance of trees and woodlands (arboriculture and tree surgery), specialist features within the garden such as walls and patios (hard landscaping) and use of water (aquatic gardening).
Twig	Shoot growth now dormant.
Variety	A group of strains or single strain of closely related plants of common origin which have similar characteristics and can be differentiated on the basis of structural and functional characters from another group of plants.
Vernalization	Method of inducing early flowering in plants by pretreatment of the propagating material with a very low temperature.
Vivipary	The phenomenon in which seed germinate in the fruit while it is still attached to the mother plant.
Water sprout	A very vigorous (unbranched) shoot arising out of adventitious buds on main scaffold branches, on the leader or in the vicinity of large pruning wounds.
Woody Perennial	A perennial that maintains live woody stem growth at the end of the growing season, e.g., Bush fruit, shrubs, trees, climbers (e.g., grape).

2

ABBREVIATIONS

Abbreviation	Full Form
AICRP	All India Co-ordinate Research Project.
2,4-D	2,4-dichlorophenoxy acetic acid.
AICVIP	All India Co-ordinate Vegetable Improvement Project.
APEDA	Agricultural and Processed Food Products Export Development Authority (New Delhi).
AVRDC	Asian Vegetable Research and Development Centre.
AVRS	Asian Vegetable Research Station.
AVT	Advanced Varietal Trial.
BBF	Broad Bed and Furrow.
BBMV	Banana Bract Mosaic Virus.
BBTV	Banana Bunchy Top Virus.
CCRS	Central Coconut Research Station.
CIHNP	Central Institute of Horticulture For Northern Plains.
CIPET	Central Institute of Post – harvest Engineering and Technology.
CISTH	Central Institute for Sub-Tropical Horticulture.
CPCRI	Central Plantation Crops Research Institute.
CPRI	Central Potato Research Institute.
CTCTI	Central Tuber Crops Research Institute.
CTV	Citrus Tristeza Virus.
CVT	Co-ordinate Varietal Trial.
DHO	District Horticulture Officer.
DPUHF	Dr. Parmar University of Horticulture and Forestry.
IAA	Indole Acetic Acid.
IIHR	Indian Institute of Horticulture Research.

Abbreviation	Full Form
IIVR	Indian Institute of Vegetable Research.
IVT	Initial Varietal Trial.
MACS	Maharashtra Association for the Cultivation of Science.
MHSCC	Maharashtra Hybrid Seeds Co. Ltd.
MSSRF	M.S. Swaminathan Research Foundation.
NAA	Naphthalene Acetic Acid.
NAFED	National Agricultural Co-operative and Marketing Federation.
NAIP	National Agricultural Innovative Project.
NARP	National Agricultural Research Project.
NATP	National Agricultural Technology Project.
NHB	National Horticulture Board.
NHRDF	National Horticulture Research and Development Foundation.
NRC	National Research Center for Citrus.
NRC	National Research Center for Grapes.
NRCO&G	National Research Center for Onion & Garlic.
NRCAH	National Research Centre on Arid Horticulture.
NRCB	National Research Center for Banana.
NRCC	National Research Center for Cashew.
NRCPC	National Research Center for Pomegranate, Solapur.
NRCTH	National Research Center for Temperate Horticulture.
PHT	Post Harvest Technology.
TSS	Total Soluble Solids.

3

REASONING

1. Why bending is done in guava?

The main object is to reduce dominance of polarity. In local varieties buds at top are sprout and the buds at lower side of shoot are failed to sprout. This behaviour of apical buds to dominate the buds situated on the lower side is termed as apical dominance of polarity.

To break this apical dominance and to bring all buds to one level and to give equal chance to every bud to sprout. The erect growing branches are bend and made horizontal, when the shoots or branches is made horizontal more auxin accumulates on the lower side of the shoot, this leads sprouting of buds. This operation is done only in erect growing varieties. 'Sardar' guava is naturally horizontal; its brandus do not required any bending.

2. Why Notching is done in Fig?

It involves partial removed of slip of bark of 2-6 mm thickness and about 15-25 mm in length. Effect of notching is interruption of downward flow of carbohydrates. Notching is done just below the bud. Season of notching is Aug-Sept.

3. Why Ringing is followed in Mango?

It is followed in mango to induce the Fruitfulness.

4. Why girdling is done in grapes?

Downward flow of carbohydrates is interrupted one week after fruit set. Girdling is done at base of cane. It is notching but removal of circular piece of bark of width of 2-5 mm below the bunch to accumulation of food material for development of bunch.

Causes of Bitterness/colour in various Fruits/Vegetables

1.	Bitterness in Almonds	Amylaedin
2.	Bitterness in Bael	Marmelotisin

3.	Rancidity in Aonla	Tannin/Ealogic/Galic/Polyphenol
4.	Yellowing in Papaya	Caricaxanthium
5.	Acridity m Colocasia	Calcium oxlate
6.	Bitterness in Bitter gourd	Momordicoside
7.	Bitterness in gourd	Cucurbitacin
8.	Bitterness in Pepper	Oleorecin
9.	Green colour in Potato tubers	Solanin
10.	Odour in Garlic (Injured)	Allicin
11.	Odour in Garlic (Uninjured)	Amino acid-Allinase
12.	Odour in onion	Allyl propyl disulphide
13.	Orange colour in Carrot	Carotene
14.	Pungency in Capsicum	Capsaicin
15.	Pungency in Radish	Inocynate
16.	Pungency in Turnip	Calcium oxlate
17.	Red colour in carrot	Anthocyanin
18.	Red colour in Chilli	Capsainthin
19.	Red colour in Onion	Anthocyanin
20.	Red colour in Tomato	Lycopene
21.	Yellow colour in Onion	Quercitin
22.	Yellow colour in Turmeric	Curcumin

S. No.	Said to be	To Whom
1.	Adam's Fig	Banana
2.	Bread fruit	*Artocarpus* sp.
3.	Bullock's heart	*Annona reticulata*
4.	Butter fruit	Avocado
5.	Egg plant	White Brinjal
6.	Elephant – Foot	Yam
7.	Kalpa Vriksha	Coconut
8.	King of Arid Fruits or poor man's fruit or Apple —— Ber	
9.	King of Flowers	Rose
10.	King of fruits	Mango

Contd...

S. No.	Said to be	To Whom
11.	King of spices	Black Pepper
12.	King of temperate fruit	Apple
13.	King of Vegetables	Potato
14.	Lady's Finger	okra
15.	Poor man's root crop	Tapioca
16.	Queen of flowers	Gladiolus
17.	Queen of fruits	Litchi
18.	Queen of pulses	Pea
19.	Queen of spices	Cardamom
20.	The tree of heaven	Coconut

4

SHORT EXPLANATIONS

Common name, botanical name and chromosome number of horticultural crops.

Common name	Botanical name	Chromosome number (2n)
Agathi	Sesbania grandiflora	24
Agave	Agave americana	58
Almond	Prunus communis	16
Amaranthus	Amaranthus spp.	32
Amaryllis	Amaryllis spp.	22
Antirrhinum	Antirrhinum majus	16
Aonla	Emblica officinalis	28
Apple	Malus domestica	34
Apricot	Prunus armeniaca	16
Areca nut	Areca catechu	32
Asparagus	Asparagus officinalis	20
Aster	Callistephus chinensis	18
Avocado	Persia americana	24
Bael	Aegle marmelos	18
Banana	Musa paradisica	22, 33, 44
Barleria	Barleria cristata	40
Beet root	Beta vulgaris	18
Ber	Zizyphus mauritiana	48 (4x)
Bitter gourd	Momordica charantia	22
Black pepper	Piper nigrum	128
Bottle gourd	Lagenaria siceraria	22
Bougainvillea	Begonia spp.	34
Brinjal	Solanum melongena	24
Brussel's sprout	Brassica oleracea	18
Bullock's heart	Annona reticulata	14
Cabbage	Brassica oleracea	18
Canna	Cannas sp.	18
Carambola	Averrhoea carambola	24
Cardamom	Elettaria cardamommum	48
Carnation	Dianthus caryophyllus	30-60
Carrot	Daucus carota	18
Cashew	Anacardium occidentale	42

Common name	Botanical name	Chromosome number (2n)
Cassava	*Manihot esculanta*	36
Cauliflower	*Brassica oleracea*	18
Celery	*Apium graveolens*	22
Chekurmanis	*Sauropus androgynous*	-
Chicory	-	18
Chilli	*Capsicum annum*	24
China Aster	*Aster chinensis*	18
Chines cabbage	*Brassica chinensis*	20
Chinese potato	*Coleus pervriflorus*	-
Chive	-	16
Chow-chow	*Sechium edule*	28
Chrysanthamum	*Dendranthema grandiflora*	18
Cinnamon	*Cinnamomum verum*	24
Clove	*Syzygium aromaticum*	44
Cluster bean	*Cyamopsis tetragonolobus*	14
Cocao	*Theobroma cocoa*	20
Coconut	*Cocus nucifera*	32
Coffee	*Coffea robusta*	22
Coriander	*Coriandrum sativum*	22
Cosmos	*Cosmos bipinnatus*	24
Cow pea	*Vigna unguiculata*	22
Cucumber	*Cucumis sativus*	14
Cumin	*Cuminum cyminum*	14
Curry leaf	*Murraya koenigi*	18
Custard apple	*Annona squamosa*	14
Dahlia	*Dahlia variabilis*	16
Date palm	*Phoenix dactylifera*	36
Dolichos bean	*Lablab purpureus*	22
Drum stick	*Moriga oleifera*	28
Elephant foot yam	*Amorphophyllus campanulatus*	26
Endive	-	18
Fenugreek	*Trigonella foenu graceum*	16
Fig	*Ficus carica*	56
French bean	*Phaseolus vulgaris*	22
Gaillardia	*Gallardia pulchella*	36
Garlic	*Allium sativum*	48
Gherkin	*Cucumis anguria*	24
Ginger	*Ginger officinalis*	22
Gladiolus	*Gallardia pulchella*	30-60
Globe artichoke	*Cynara scolymus*	34
Grape	*Vitis vinifera*	38
Grape fruit	*Citrus paradisi*	18
Greater yam	*Dioscorea alata*	40

Common name	Botanical name	Chromosome number (2n)
Guava	*Psidium guajava*	22
Ivy gourd	*Coccinia indica*	24
Jack fruit	*Artocarpus heterophyllus*	56
Jamun	*Syzygium cumunii*	40
Jerusalem artichoke	*Helianthus tuberosus*	102
Kagzi lime	*Citrus aurantifolia*	18
Kale	*Brassica oleracea*	18
Karonda	*Carrisa carandas*	22
Kiwi fruit	*Actinidia deliciosa*	58
Knol-Khol	*Brassica oleracea*	18
Kokum	*Garcinia indica*	24
Leaf mustard	*Brassica juncea*	36
Leek	*Allium porum*	32
Lemon	*Citrus limon*	18
Lesser yam	*Dioscorea esculenta*	40
Lettuce	*Lactuca sativa*	18
Lima bean	*Phaseolus lunatus*	-
Litchi	*Litchi chinensis*	30
Long melon	*Cucumis melo*	24
Loquat	*Eriobotrya japonica*	34
Lotus	*Nelumbo nucifera*	16
Macadamia nut	*Macadamia temifolia*	48
Mandarin	*Citrus reticulata*	18
Mango	*Mangifera indica*	40
Mangosteen	*Garcinia mangosteena*	24
Marigold	*Rose* spp.	24-48
Musk melon	*Cucumis melo*	24
Narcissus	*Narcissus* spp.	14
Oil palm	*Ealias guinensis*	32
Okra	*Abelmoschus esculantus*	130
Onion	*Allium cepa*	16
Palak	*Beta vulgaris*	18
Palmyra	*Borasus flaballifer*	32
Papaya	*Carica papaya*	18
Passion fruit	*Passiflora edulis*	18
Peach	*Prunus persia*	16
Pear	*Pyrus communis*	34
Peas	*Pisum sativum*	14
Pecanut	*Carya illieonsis*	36
Persimmon	*Diospyras kaki*	90
Phalsa	*Grewia subinequalis*	36
Pineapple	*Annanas comosus*	50
Pistachio nut	*Pistachia vera*	30
Plum	*Prunus domestica*	16 (Japanese) 48 (European)

Common name	Botanical name	Chromosome number (2n)
Pointed gourd	*Trichosanthus dioca*	22
Pomegranate	*Punica granatum*	18
Potato	*Solanum tuberosum*	48
Pumkin	*Cucurbita moschata*	40
Radish	*Raphanus sativus*	18
Ridge gourd	*Luffa acutangula*	26
Rose	*Rosa indica*	14
Round melon	*Citrullus lanatus*	22
Salvia	*Salvia splendens*	12
Sapota	*Achrus zapota or Manilkara achras*	26
Shallot	*Allium ascalonicum*	16
Snake gourd	*Trichosanthus anguina*	22
Snapmelon	*Cucumis melo*	24
Spinach	*Spinacea oleracea*	12
Spine gourd	*Momordica cochinchinesis*	28
Sponge gourd	*Luffa cylindrica*	26
Sprouting Broccoli	*Brassica oleracea*	18
Straw berry	*Fragaria ananasa*	56
Summer squash	*Cucurbita pepo*	40
Sweet cherry	*Prunus avium*	16
Sweet corn	*Zea mays*	20
Sweet orange	*Citrus sinensis*	18
Sweet pepper	*Capsicum annum*	24
Sweet potato	*Ipomea batatas*	90
Tamarind	*Tamarindus indica*	24
Tannia	*Xanthosoma sagittifolium*	26
Taro	*Colocasia esculenta*	28
Tea	*Camelia sinensis*	30
Tomato	*Lycopersicon esculentum*	24
Tuberose	*Polianthus tuberosa*	60
Tulip	*Tuplica stellata*	24-60
Turmeric	*Curcuma longa*	62
Turnip	*Brassica rapa*	20
Verbena	*Verbena hybrida*	10
Walnut	*Juglans regia*	32
Water melon	*Citrullus lanatus*	22
Wax gourd	*Benincasa hispida*	24
White yam	*Dioscorea rotundata*	40
Winged bean	*Psophocarpus tetragonolobus*	18
Winter squash	*Cucurbita maxima*	40
Wood apple	*Feronia limonica*	18
Zinnia	*Zinnia elegans*	24

Fruit crop and their center of origin

Common name	Origin of the crop
Almond	Afghanistan
Aonla	Indo-China
Apple	Asia minor to Western Himalayas
Apricot	China
Areca nut	
Avocado	Mexico
Bael	India
Banana	South-east Asia
Ber	
Bullock's heart	
Carambola	Indo-Malaya
Cashew nut	Brazil
Cocao	
Coconut	South-east Asia
Coffee	
Custard apple	Tropical America (Mexico)
Date palm	West-Asia (Iraq)
Fig	West-Asia (Iraq)
Grape	
Grape fruit	
Guava	Peru
Jack fruit	India
Jamun	Indo-Malaya
Kagzi lime	India
Karonda	
Kiwi fruit	
Lemon	South-east Asia
Litchi	China
Loquat	China
Macadamia nut	
Mandarin	China
Mango	Indo-Burma
Mangosteen	
Oil palm	
Palmyra	
Papaya	Tropical America (Mexico)
Passion fruit	Brazil
Peach	China
Pear	Europe
Pecanut	USA

Common name	Origin of the crop
Persimmon	China
Phalsa	India
Pineapple	Brazil
Pistachio nut	
Plum	Japan
Pomegranate	
Sapota	Mexico
Straw berry	Man made hybrid
Sweet cherry	
Sweet orange	China
Tamarind	
Tea	
Walnut	Indo-China
Wood apple	

Horticultural crops and their improved cultivars.

Crop	Improved cultivars
Almond	Sole, Neplus ultra, Texas, Merced, Peerless, Drake, Katha, Dhebar, Makhdoom, Non-pareil, IXL 7, Jordanolo, etc.
Amaranthus	Arka Kiran, Arka Saguna, Chhoti Chawli, Pusa Kirti, Pusa Lal Chawli, Badi Chawli etc.
Aonla	Kanchan (NA-4), Banarasi, Francis (Hathijhool), Chakiya, NA-5 (Krishna), NA-6, NA-7 (Amrit), NA-9 (Neelum), etc.
Apple	Irish peach, Benoni, Fenny, Tydemans early, Early shanburry, Red gold, Lord lambourne, Top red, Red delicious, Richard, American mother, Rome beauty, Razakwar, Mcintosh, Jonathan, Golden delicious, Cortland, Red chief, super type, Green English vars, Lal Ambri, Amb Red, Ambroyol, Amrich, Sunheri, Chaubattia anupam, etc.
Apricot	Khante, New castle, Moonrpark, Alfred, Nugget, Farmingdale, Bebeco, Royal, St. Ambroise, Chaubattria Alankar, Early Shipley, Halman, Charmagz, Kaisha, Chaubattria Madhu, Chaubattria Kesari, etc.
Areca nut	Sreevardhani, Sumangla, Mangla, Sree mangla, Samrudhi, Mohitnagar, etc.
Ash gourd	Pusa Ujjwal, Mudlier etc.
Avocado	Green (Guatemalan type), Purple (West Indian type), Furete, etc.,
Bael	Kagzi Gonda, Kagzi Banarasi, Kagzi Etawah, etc.
Banana	Robusta, Dwarf Cavendish, Lady finger, Grand Naine, Amritsagar, Poovan, Pey kunnan, Poovan mysore, Ney poovan, Lal velchi, Monthan, Nendran, Hill banana, Bodles altafort, FHIA-1, CO-1, Klue teparod, Rajapuri, Nendran, Moongli etc.
Beet root	Crosby Egyptian, Crimson Globe, Early wonder, Detroit red etc.
Ber	Gola, Umaran, Illaichi, Seb, Meharun, Rashmi, Mundia, Banarasi, Chohara, Kadaka, Chanur, Mukta etc.
Betel vine	Mitha, Kallipatti, Sanchi, Karpoori, Bangla, etc.

Crop	Improved cultivars
Bitter gourd	Konkan Tara, Pusa Vishesh, Preethi, Harkani, Priyanka, Coimbatore long, Pusa-Do-Mausami, Phule Green Gold, Pusa hybrid-1 Hirkani, Phule Green Gold, Phule priyanka, Phule Ujwala, Konkan tara etc.
Black pepper	Sreekara, Poomima, Karimunda, Panchami, Subhakara, Panniyur-2, 4, 6 etc.
Bottle gourd	Punjab long, Arka Bahar, Pusa Summer Prolific round, Samarat, Punjab Round, Pusa Round, Pusa Naveen, Pusa Summer, Harita, Gutka, Pusa Meghdoot, Pusa Mangari, Punjab Komal, Varad, Pusa Sandesh, PBOG-1, NDBG-1, etc.
Brinjal	Pusa purple long, Pusa purple cluster, Pant samrat, Pusa purple round, Punjab chamkila, Arka Nidhi, Arka kusumkar, Punjab bahar, Azad kranti, Punjab Neelum, Arka Shirish, Punjab barsati, Pusa bindu, Hissar Shyamal, Hissar Jamuni, Pusa upkar, Krishna, Manjarigota, Pragati, Vaishali, Phule Harit Anuradha, ABV 1 etc.
Broad bean	Red Epicure, Windsore type, Pusa Sumeet etc.
Broccoli	Palam Samridhi, Green Mountain, Green Head, Decicco, Coastal Atlantic, Sparten early, Italian Green, Greenbud, Pusa KTS-1 etc.
Cabbage	Golden Acre, Red Acre, August early variety, Copenhagen market, September, Glory of Enkhuizen, Pride of India, Pusa Ageti, Pusa synthetic, Pusa Sambandh, Pusa Mukta, Pusa Drumhead, Questo, Sree Ganesh Gol, Uttam, Bajrang, Green Boy and green express, Sudha, Stone head etc.
Carambola	Gold star, Golden star, Icambola, etc.
Cardamom	PV-2, Mysore type, Malabar type, Mudigree-1, Vazukha type, Bebo, Golsey, Ramsey, Ramla etc.
Carrot	Pusa Kesar, Pusa Meghali, Pusa Yamdagini, Chantenay, Zeno, Danvers, Nantes half long, Early nantes etc.
Cashew nut	Vengurla-1 to 7, Dhana, Madakhathara, Priyanka-export variety, Ullal-1, Chintamani-1, Ullal-2, Ullal-3, UN-50, Ven-3, Ven-4, Ven-6, VRT-1 to 3, etc.
Cauliflower	Improved Japanese, Pusa Himjyoti, Pusa Snowball K-1, Pusa Ketki, Pusa Aghani, Pusa Deepali, Pusa Shubhra, Pusa early synthetic, Pant Gobi-3, White fleash, Early Himlata, Nath Ujwala, Nath Shweta, Himani, Cash more, Candid charm, Pusa Sharad, Pusa Snoball etc.
Celery	Florida golden, Golden, Wrights grove giant, Ford hook emperor etc.
Cherry	Sam, white heart, Black republican, Sunnit, Pink early, Compact stella, Lambert, Sunbrust, Stella, Governors wood, Napolean white, Black heart, Early rivers, etc.
Chilli	Sweet Banana, Bullnose, Chinese Giant, Yolo wonder, California wonder, World beater, Golden wonder, Arka Mohini, Arka Gaurav, Arka Basant, Pusa Deepti, Green Gold, Bharat, Lario, Hira, India, Early Bounty, Pusa Meghdt, Phule Mukta, Phule Saryamukhi, Phule Sai, Phule Jyoti, Musalwadi, Aynirekha, Parbhani Tejes, Konkan Kirti, Surkta, Jayanti etc.
Chow-chow	Chayote, Choco, Askas etc.
Cinnamon	Navashree, Yercaud-1, Nithyashree, Konkan Tej etc.
Clove	Pevang, Lenzipore, Amulya etc.
Cluster bean	Pusa Sadabahar, Pusa Mausmi, Sharad Bahar, Pusa Navbahar etc.
Cocoa	Trinitarion, Criollo, Forestero, etc.

Crop	Improved cultivars
Coconut	West coast tall, Sanramon, Pratap, Laguna, Laccadive ordinary, East west tall, Andaman ordinary, Nuleka, Mangipod, Chowghat orange dwarf, Gangabondam, Gudanjali, Chowghat green dwarf, Coco Nino, Lakshadeep, Ordinary, TxD, Philipines, DxT, Pratap, Banavali etc.
Coffee	S-795, C x R, Agro, San ramon, Cioccie, Kent, Cauvery, Chicks, Blue mountain, etc.
Coriander	Sadhana, Pant Harithna, Rajendra Swathi, Sindhu, Swathi, Karan, Merrocan etc.
Cowpea	Arka Suman, Arka Samrudhi, Yard long bean, Arka Garin, Pusa Barsati, Phillipines early, Pusa phaslguni, Pusa Dofasli, Pusa Rituraj, Pusa Komal etc.
Cucumber	China long, Staright eight, Poinsett, Japanese long Green, Sheetal, Himangi, Phule subhangi, Pusa Sayog, Priya, etc.
Cumin	RZ-19, GC-2, RZ-209, Vijapur-5, GC-1 etc.
Curry leaf	Suwasani
Custard apple	Barbados seedling, British Guinea, Washington, Mahaboobnagar, Balanagar, Kakarlapahad, African pride, Arka sahan, TP 7, etc.
Date palm	Sharan, Zahidi, Fresh eating, Chuhharah making, Bread type date, Cane sugar date, Invert sugar date, etc.
Dolichous bean	Pusa early Rajni, Pusa early prolific, Arka Jay, Konkan Bhushan, Rajni, Deepali, CO-10 etc.
Drum stick	Jaffna, PKM-1, PKM-2 etc.
Eucalyptus	Yeshwant
Fennel	RF-101, RF-125, RF-35 etc.
Fenugreek	Hissar Sonali, Rajendra Kranti etc.
Fig	Banglore, Lucknow, Merselies, Black ischia, Kabul, Dinkar, Puna fig, Denna, Dinkar etc.
French bean	Tweed wonder, Pusa Himlata, Kentuky wonder, Contender, Top cross, Bountiful, Arka suvidha, Giant Stringless, Arka Komal, Pusa Parvati, Pant Anupama, Jampa, etc.
Garlic	Sweta, Yamuna Safed, G-282, Godavari, Agrifound Parvati, Agrifound white (G-41), Godawari, Sweta etc.
Ginger	Surbhi, Suphrabha, Suruchi, Wynad manantody, Himgiri, Riode Janerio etc.
Grape	Thompson seedless, Banglore blue, Gulabi, Kishmish chorni, Beauty seedless, Sharad seedless, Anab-a-shahi, Dilkhush, Perlette, Pusa seedless, Tas-A-Ganesh, Sonnaka, Manik chaman, Arkavati, Arka neelmani, Arka Krishna, Arka hans, Arka shyam, Arka trishna, Arka shweta, Arka chitrah, Pusa urvashi, Pusa navrang, Arka kanchan, Bhokri, Cheema Sahebi etc.
Grape fruit	Foster, Marsh seedless, Triumph, Red blush, Thompson, Sharanpur special, Duncan, etc.
Guava	Allahabad safed, Lucknow-49 (Sardar), Hafsi, Chittidar, Harijiha, Allahaba surkha, Nagpur seedless, Arka Amulya, Saharanpur seedless, Behat coconut, Arka nridula, Hissar anulya, Apple clour, Lalit, Kohir safed, Safed jam, etc.
Jackfruit	Champa, Hazari, Muttam varikka, Gulabi, Monkey jack, Rudarkshi jack, Singapore or Ceylon jack etc.

Crop	Improved cultivars
Jamun	Raj Jamun, Paras (large size), Narendra Jamun-6 (seed less)
Kagzi lime	Vikram, Chakradhur, PKM-1, Jai devi, Sai sarbati, Pramalini, etc.
Karonda	Maroon (Green, white and dark purple fruits, etc.
Kiwifruit	Allison, Bruno, Monty, Abbott, Tomuri, etc.
Knol-khol	White Vienna, Purple Vienna, King of North, Golithwhite, Sutton's early purple, Purple speck etc.
Kokum	Konkan Amrit
Lemon	Lisbon, Kagzikalam, Nepali oblong, Nepali round, Villafrance, Pant lemon, Eureka, Lucknow seedless, etc.
Lettuce	Slobott, Chinese yellow, White Boston, Dark Green, Great lakes, Imperial-859, Punjab lettuce No.1 etc.
Lima bean	Challenger, Florida butter, Karolina butter, King of Garden, Baby potato, baby fordhook, Handerson bush, Wilbur, Hopi etc.
Lime and Lemons	Tahiti lime, Rangpur lime, Sweet lime, Pummelo, Kagzilime, etc.
Litchi	Rose scented, Dehradun, Calcutta, Purbi, Early seedless, Gulabi, Shahi, Late seedless, Swaran roopa, Kasba, Desi, Bombai, China, etc.
Loquat	Safed, Matchless, Fire ball, Large agra, Mammoth, Thames pride, Pale yellow, Golden yellow, Tanaka, California advance, etc.
Mandarin orange	Khasi, Coorg, Emperor, Fuetrelles, Nagpur, Satsuma, Laddu, Sutwal, Kinnow, etc.
Mango	Alphonso, Banganpalli, Bombay Green, Chausa, Dashehari, Fazli, Kesar, Langra, Niranjan, Neelum, Rosica, Madhulica, Lal sindhuri, Gulab khas, Totapuri, Pairi, Banglora, Rumani, Himasagar, Kishanbhog, Mankurad, Mallika, Amrapalli, Prabha sankar, Manjeera, Ratna, Arka puneet, Sindhu, Arka aruna, Arka neelkiran, Akshay, Pusa arumina, Pusa surya Rahuri Sugandh, Niranjan, Parbhani Bhushan, Ratna, Shindu, Konkan ruche, Hapus etc.
Mangosteen	Jolo
Muskmelon	Pusa Madhuras, Hara Madhu, Arka Jeet, Arka Rajhans, Durga pura Madhu, Punjab Rasila, Hissar Madhur, Punjab Sunheri, Pusa Sharbati, Pusa Rasraj, Shweta, Swarna etc.
Okra	Pusa Mukhmali, Gujarat Behdi No.1, Pusa Sawani, Punjab Padmini, Parbhani Kranti, Arka Anamica, Arka Abhay, Panchali, Adhumik, Suriya, Varsha, EMS-8, Arka Varsha Upkar, Hissar Barsati, Pusa-A-4, Phule Kirti, Phule Utkarsha, Parbhani Kranti etc.
Onion	Early Grano, Bermuda Yellow, Brown Spanish, Pusa White Round, Pusa Ratnar, Pusa Red, N-53, Arka Kalyan, Arka Bindu, Agrifound Dark Red, Banglore Rose, Pant Red, Nashik Red, Arka Niketan, Pusa White Flat Pusa Madhvi, Arka Pitambar, Arka Kirtiman, Arka Lalima, Phule Samarth, Phule Suvarna, Phule safed, N 2-4-1, B 780, N 257-9-1 etc.
Papaya	Pusa majesty, Pusa delicious, CO-3, Taiwan, Surya, Coorg honey dew, Sunrise solo, Pusa giant, Pusa dwarf, CO-1, CO-2, CO-5, CO-6, Pant-C-1, Betty, Sunny bank, Hatras gold, CO-3, CO-7, CO-4, etc.
Passion fruit	Novel's special
Pea	Early Badger, Early superb, Little marvel, Sylvia, Lincoln, Bonneville, Arkel, Meteoror, Perfection new line, Arka Ajit, Asauji, Harbhajan, Pant upkar, etc.

Crop	Improved cultivars
Peach	Alton, World's earliest, Red haven, Stark red gold, Early candor, Saharanpur prabhat, Quetta, Julie Elberta, Alexander, Co smith, etc.
Pear	Anjou, Max red Bartlett, Bartlett, Winter Nellis, Dr. Jule's Guyot, Fertility, Laxton's superb, Flemish beauty, Starkimson delicious, Comice, Kiffer, Leconte, Patharanakh, Gola, China pear, Red blush, Punjab gold, Punjab beauty, Punjab nectar, Magness, etc.
Pecanut	Mahan, Wichita, Stuart, Nellis, Burkett, Chicksaw, Western, Desirable, Cheyenne, etc.
Persimon	Fuyu, Jiro, Mastumoto, Hiratanenashi, Hachiya, Nightingale, 20th Century, Triumph, etc.
Phalsa	Sharbati, Local, etc.
Plum	Methley, Kelsey, Settler, Beauty, Early subza, Santa Rosa, Cloth of Gold, Frontier, Victoria, Burbank, Mariposa, Red Ace, Satsuma, Elephant heart, Silver wilkson, Grand Duke, Late yellow, Prune, Green Gage, Alucha purple, Titron, Alubukhara, Satluj purple, Golden drop, Stanley, Yellow egg, President, Grand duke, Diamond, Tragedy, etc.
Pointed gourd	Swarna Rekha, Shankolia, Dandli, Damodar, Chhota Hilli, Swarna Alaukik, Kalyani, Shankolia, Bihar Sherif etc.
Pomegranate	Muskati red, Spanish ruby, Paper shelled, Karadi, Muskati, Dholka, Nabha, Alandi, Madhugiri, Chawla, Country large red, Jalore seedless, Ganesh, P 26, Mridula, Jyoti, G-137, Bassein seedless, P 23, P 26, G 137, Mridula, Phule Arakta, Bhagwa etc.
Potato	Kufri Chandramukhi, Kufri Ashoka, Kufri Lavkar, Kufri Jawahar, Kufri Deva, Kufri Megha, Kufri Jyoti Kufri Kufri Swarna, Kufri Badshah, Kufri Giriraj, Kufri Sindhri, Kufri Lalima, Kufri Chipsoma, Kufri Sutlej, Kufri Pukhraj, Kufri Red, Kufri Safed, Kufri Moti, Kufri Kundan, Kufri Naveen, Kufri Jeevan, Kufri Kuber, Kufri Megha, Kufri Ashoka, Kufri Alankar, Kufri Anand, etc.
Pumpkin	Pusa Vikas, Pusa Viswas, Ambili, Arka Surya Mukhi, Arka Chandan etc.
Radish	Pusa Desi, Pusa Chetki, Pusa Rashmi, Newari, Punjab Safed, Arka Nishant, Japanese White, Chinese pink, Scarlet Icicle, Scarlet Globe, Pusa Himani, Rapid Red etc.
Ridge gourd	Konkan Harita, Punjab Sadabahar, Pusa Nasdar, Satputia, Surekha, Arka Sujat, Phule Sucheta, Konkan Harita etc.
Sapota	Cricket ball, Kirti bharti, Kalipatti, CO-2, PKM-1, Murrabba, Guthi, Baramasi, Chhatri, Pala, Oval, Calcutta special round, CO-1, DSH-2, DSH-1, PKM-3, PKM-2, etc.
Snake gourd	Konkan Sheweta, Arka Suryamukhi, Pink Banana, Kaddu, Phule Vaibhav etc.
Spinach	Pusa Bharti, Pusa Jyoti, Punjab Green, Pusa Palak, Pusa Harit, Banerjee's Giant, Jobner Green etc.
Sponge gourd	Pusa Supriya, Harita, Phule Prajakata, Pusa Chikni, Phule Prajakta etc.
Strawberry	Selva, Chandler, Tioga, Pajaro, Premier, Torrey, Red coat, Fern, Dilpasand, Belrubi, etc.
Summer squash	Patty Pan, Austrian Green, Early yellow prolific, Pusa Alankar etc.,
Sweet lime	Mithotra, Mitha chikni, etc.
Sweet orange	Jaffa, Valencia, Hamlin, Pineapple, Washington navel, Batavin, Shamouti, Mosambi, Satagudi, Blood red, Mudkhed, etc.

Crop	Improved cultivars
Sweet potato	Samrat, Pusa Lal, Gouri, Kalmegh, Kiran, Shankar, Varsha, Sree Vardhani, Sree Nandini, Pusa Sunderi, Bhuban, Pusa Suffaida, Sree Bhadra etc.
Tamarind	Urigam
Tea	Sundaram, Jayaram, Golconda, Brooklands, Athrey, Singara, etc.
Tomato	Labonita, Marvel, Moneymaker, Ageti, Tiptop. Best of all, Roma, Pant Bahar, Sioux, Sonali, Arka Vikas, Arka Saurabh, Pusa ruby, Pusa Gaurav, Marglobe, Sweet-72, Pusa Sheetal, Hissar Lalima, Hissar Latit, Bhagyashree, Dhanashree, Rajashree, Phule Raja, ATV 1, Devgiri, ATH 1, Sonali, Vusundhara Parbhani, Yashree etc.
Tubrose	Phule Rajni
Turmeric	Suroma, Rajendra Sonia, Prabha, Kranti, Rashmi, Pratibha, Krishna, Srobha, Sugandham, Allepy turmeric, Suvarna, Saguna, Roma, Phule Swarupa etc.
Turnip	Pusa kanchan, Pusa Swati, Punjab Safed, Pusa Chandrima, Golden Bold, Pusa Swarnima etc.
Walnut	Lake English, Placentia, Roopa, Eureka, Franquette, Karan, Wilson, Gobind, Chakrata, etc.
Water melon	Sugar baby, Dixie cream, New Hemisphire midget, Furken, Improved Shipper, Asahi Yamato, Durgapura Meetha, Durgapura Kesar, Pusa Rasal, Arka Manik, Arka Jyoti, Pusa Bedana, Madhur, Milan etc.
Wood apple	Kowath, Kainth, Ellora etc.

Horticultural Institutions Run by ICAR

Abbreviation	Full Form	Place of Center
CCRI	Central Coffee Research Institute	Chikmagalur
CFQC&TI	Central Fertilizer Quality Control and Training Institute	Faridabad
CIAH	Central Institute for Arid Horticulture	Bikaner
CIPHET	Central Research Institute for Post Harvest Engineering and Technology	Ludhiana
CISTH	Central Institute for Subtropical Horticulture	Lucknow
CITH	Central Institute for Temperate Horticulture	Srinagar
CPPTI	Central Plant Protection Training Institute	Hyderabad
CPRI	Central Potato Research Institute	Shimla
CRIC	Central Research Institute for Chikoo	Muzaffarpur
CSSRI	Central Soil Salinity Research Institute	Karnal
CSWCRIT	Central Soil and Water Conservation Research and Training Institute	Dehradun
CTCRI	Central Tuber Crops Research Institute Tiruvanantapuram	
DHIHQ	Division of Horticulture at ICAR Headquarter	New Delhi
IARC	Internet Agriculture Research Center	
IIHR	Indian Institute of Horticulture Research	Bangalore
IISR	Indian Institute of Spice Research	Calicut
IIVR	Indian Institute of Vegetable Research	Varanasi
NAIS	National Agriculture Insurance Scheme	

Abbreviation	Full form	Place of Center
NARP	National Agriculture Research Project	
NARS	National Agriculture Research System	
NATP	National Agriculture Research Center	
NIAM	National Institute of Agriculture Marketing	Jaipur
NPOP	National Programme for Organic Production	
NPPTI	National Plant Protection Training Institute	Hyderabad
NRCA	National Research Center for Agro Forestry	Jhansi
NRCB	National Research Center for Banana	Tiruchirapalli
NRCC	National Research Center for Cashew	Puttur
NRCC	National Research Center for Citrus	Nagpur
NRCG	National Research Center for Grapes	Pune
NRCG	National Research Center for Groundnut	Junagadh
NRCIPM	National Research Center for Integrated	New Delhi
NRCL	National Research Center for Litchi	Muzaffarpur
NRCM	National Research Center for Mushroom	Solan
NRCM	National Research Center for Mustard	Bharatpur
NRCM&P	National Research Center for Medicinal and Aromatic Plants	Anand
NRCO	National Research Center for Orchids	Gangtok
NRCO&G	National Research Center for Onion	Nasik
NRCOP	National Research Center for oil Palm	Eluru
NRCPB	National Research Center for Plant Biotechnology	New Delhi
NRCS	National Research Center for Sorghum	Hyderabad
NRCS	National Research Center for Soybean	Indore
NRCSS	National Research Center for Seed Spices	Ajmer
NRCW	National Research Center for Weed Science	Jabalpur
PDBC	Project Directorates on Biological Control	Bangalore
PDOS	Project Directorates Oil Seed	Hyderabad
PDP	Project Directorates on Pulses	Kanpur
PDR	Project Directorates on Rice	Hyderabad
PDVR	Project Directorates on Vegetable Research	Varanasi
PDW	Project Directorates Wheat	Karnal
PDWM	Project Directorates on Water Management	Rahuri
RRII	Rubber Research Institute of India	Kottyayam

1. What is Bahar treatment ?

Bahar treatment means withholding of water one to one and half months before flowering with a view to give rest to the tree. It leads to more carbohydrate accumulation. The choice of bahar depends upon availability of water, market and incidence of diseases & pests. There are three bahars taken in fruits crops.

I.   Ambia bahar — Jan – Feb

II. Mrig bahar — June – July

III. Hasta bahar — Sept – Oct.

2. What is training ?

When a plant tied or supported over a trellies or Pergola in certain fashion or some of its parts are cut off with a view of giving the plant better frame work, the operation is called as training systems.

There one three training systems :

(1) **Central leader system :** Trees are trained to form a trunk extend from ground to the top of the trees, e.g., peach.

(2) **Open centre system :** The main stem is allowed to grow only up to a certain height and is headed back with in a year of planning of tree, e.g., apple, plums.

(3) **Modified leader system :** Intermediate between central leader and open centre system, e.g., guava, fig.

3. What is pruning?

It is the art of removing certain portion of the plant with a view to divert the sap flow towards the fruiting area present on the plant to produce more and superior quality of fruits.

There are two methods of pruning :

(1) **Heading back :** The terminal portion of every twig, canes and shoots are cut, retaining basal potion, It is practiced in hedges, ornamentals, grapes etc.

(2) **Thinning out :** Some twigs, canes or shoots are removed from their point of origin and retaining others. It is practiced in fruit trees.

4. What are the types of vegetable gardens?

Vegetable gardens divided into seven types according to purpose.

(a) **Kitchen garden or Home garden :** The main purpose is to provide the family daily with fresh vegetables rich in nutrients and energy.

(b) **Market garden :** The main purpose is to supply the vegetables for the local market.

(c) **Truck garden :** In which produces special crops in the large quantity for distant market *e.g.* potato , onion.

(d) **Vegetable forcing :** In which the vegetables are produced out of their normal season, *e.g.,* in glass house.

(e) **Vegetable garden for processing :** In this preservation of vegetable for consuming them in unavailable seasons.

(f) **Vegetable garden for seed production :** Which are concerned with the good seed production.

(g) **Floating vegetable garden :** This types are seen on the Dal lake of the Kashmir valley. Floating base is prepared and seedlings are transplanted.

5. What are the advantages of growing horticulture crops?

India has a wide variety of climate and soils on which a large number of horticultural crops such as fruits, vegetables, potato, tropical tuber crops, mushroom, ornamental, medicinal and aromatic plants, plantation crops, spices, cashew, cocoa and betelvine are grown. Diversification in horticulture is a best option as there are several advantages of growing horticultural crops. These crops :

- produce higher biomass than field crops per unit area resulting in efficient utilization of natural resources.
- Are highly remunerative for replacing subsistence farming and thus elevate poverty level in rainfed, dryland, hilly, arid and coastal agro-ecosystems.
- Have potential for development of wastelands through planned strategies.
- Need comparatively less water than food crops.
- Provide higher employment opportunity.
- Are important for nutritional security.
- Are environment – friendly.
- Are high-value crops with high potential of value-addition.
- Have high potential for foreign exchange earnings.
- Make higher ontribution to GDP.

6. What are the research infrastructure in horticulture exists in the country?

At present, following research infrastructures in horticulture exists in the country under the National Agriculture Research System (NARS).

- 10 central institutes with 27 regional stations.
- 10 National Research Centres on important crops.
- 9 multidisciplinary Institutes working on horticulture crops.

- 15 All India co-ordinated Research projects with 215 centers.
- One state Agricultural university in horticulture and Forestry.
- 25 State Agricultural University with Horticulture discipline.
- 5 network projects.
- 330 adhoc projects.
- 29 revolving fund projects/Schemes.

1n addition, a number of institutes under CSIR and Ministry of Commerce are also devoted to research on ornamental, medicinal and aromatic crops besides on aspects of post harvest technology.

7. Which are the planting systems for fruit Crops ?

- Square system
- Rectangular system
- Triangular system
- Hexagonal system
- Quincunx system (diagonal system)
- Contour system
- Terrace system.

8. What is AVRDC?

Asian Vegetable Research and Development Centre Taiwan, was established in 1971, with main objective was to improve production and marketing of vegetable crops in the lowland tropics. It is an international institute for vegetable research, development and training.

9. Classification of Horticultural Crops

Why to done classification of Horticultural Crops?

- Easy cultural and agronomical operation
- For breeding purpose
- For propagation use
- For plant protection purpose
- Suitability to different climatic conditions.

Horticultural crops are popularly classified into the three broad divisions of fruits, vegetables and flowers. The Horticultural crops are also classified on different basis.

I. According to growth habit & physiological characters:

(a) Herbs – lawn grosses, Ageratum

(b) Shrubs – Nerium, Hibiscus

(c) Climbers – bouganvelia

(d) Creepars – bigonia

(e) Rees – Tamarind, mango, raintree.

II. According to whether they shed their leaves during a year

(a) Deciduous – Fig, phalsa, apple, Guava, ber, pomegranate

(b) Evergreen – Mango coconut, banana, k lime.

III. According to seasonal basis

(a) Summer Season – Cucurbits, okra, clusterbean

(b) Rainy season – Chilli, brinjal, to mato, onion

(c) Winter season – cole crops, peas

IV. According to life span of crop

(a) Annuals – flowers

(b) Biennials – carrot, onion

(c) Perennials – durmstic, little gourd

V. According to their temperature requirements.

(a) Temperate – apple, plum, grape

(b) Tropical – Guava, grape

(c) Tropical – banana, citrus, sapota

VI. According to use of plants :-

(a) Edible plants, (b) Ornamental plants

Edible Plants are again classified as (i) Vegetables, (ii) Fruits

(i) **Vegetables** — (A) Vegetables grown for the aerial portion

(i) Cole Crops—cabbage, Knolkhol cauliflower

(ii) Legume Crops — Peas, beans

(iii) Solanaceous veg. — brinjal, chilli, tomato

(iv) cucurbits — cucumber, bitter gourd, bottle goard.

(v) Leafy veg. — palak , methi

(vi) Salad veg. — lettuce, broccoli

(B)   Vegetables grown for under ground portions

    (i)   Root veg. — carrot, radish

    (ii)   Tuber veg — Potato, yam

    (iii)   Bulb veg — onion, garlic

(ii)   **Fruits** – (A) Temperate fruits

    (i)   Small fruits — Strawberry

    (ii)   Tree fruits — apple, pear, fig

    (iii)   Nuts — Peanut, walnut.

(B)   Tropical & Subtropical fruits

    (i)   Herbaceous perennials -Pineapple, banana

    (ii)   Tree fruits -Mango, papaya

    (iii)   Nuts -Coconut, Arecanut, cashewnut

(b)   Ornamental plants

    (i)   Flowering trees — Gulmohar, Neelmohar, cassia

    (ii)   Road side trees — Rain tree, Neem, baniyan

    (iii)   Shade giving trees — Rain tree, mohogani

    (iv)   Flowering shrubs — Hibiscus , Nerium, Tagar

    (v)   Foliage shrubs — Cushurina , Thuja

    (vi)   Climbers & Creepers — Begonia , Ipomea , Petrea

    (vii)   Bulbous plants — Canna , tuberose.

    (viii)   Hedges & Edges — Duranta , cleredendron

    (ix)   Annuals — Petunia, zinnia

    (x)   Perennials — Roses, chrysanthemum.

Methods of Plant Propagation

  I.   **Sexual methods** : Use of seed

(A)   Requiring preparatory treatments

    (i)   Chemical — Asparagus

    (ii)   Mechanical — Canna , lotus

    (iii)   Scalding (pouring boiling water) — Coffee, chiku

    (iv)   Soaking — Peas, rain tree

    (v)   Stratification (moist chilling-low temp) — Peach, plum

(B) Requiring No treatments — annuals & perennials

II. Asexual methods :-use of stem, root, leaf, etc.

(A) Common origin methods (3)

   (i) Plants parts generally detached before rooting

      (a) Separation — Bulbs – lilies

          — Corms – Gladiolus

      (b) division — Rhizomes – Canna, ginger

          — Offsets – pineapple

          — Tubers – potato, dahalia

          — Suckers – banana

          — Runners – Hariyali

   (ii) Plants park invariably detached before rooting (cuttage)

      (a) Stem cutting — Hard wood cutting – Grape, rose

          — Semi hard wood cutting – panax, Arelia

          — Soft wood cutting – Geranium, tomato.

      (b) Leaf cutting — use of petiole – Pepromia

          — use of marginal buds – bryophyllum

          — use of notches & main veins – begonia

      (c) Root cutting — guava , bread fruits

   (iii) Plant parts invariably not detached before rooting (Layerage)

      (a) Air layer or gottee – mango, sapota

      (b) Air layer with plastic – guava , pomegranate

      (c) Tongue layering – Guava

      (d) Tip layering – Raspberry

      (e) Compound layering – Rose

      (f) Trench layering – plum, apple

(B) Separate Origin methods (2)

   (1) Grafting — use of bud stick

(A) Scion attached methods

   (i) Simple approach or Inarching – Mango

   (ii) Tongue grafting – Sapota

   (iii) Saddle grafting – jackfruits

(B) Scion detached methods

    (i)   Veneer grafting

    (ii) Saddle grafting

    (iii) Wedge grafting

    (iv) whip grafting

    (v)   Whip & tongue grafting

(C) Methods of grafting on established trees.

    (i)   Side grafting – mango

    (ii) Crown grafting –

    (iii) Top working by

        (a) Inarching

        (b) by crown grafting

        (c) by forkert budding.

(D) Renovation methods

    (i)   Bridge grafting

    (ii) Buttress grafting

(2) Budding — use of single bad

    (a)   Shield budding — peach

    (b)   'T' '4%' and 'I' budding – citrus

    (c)   Ring budding — ber

    (d)   Patch budding — cashewnut

    (e)   Forkert budding — mango

    (f)   Flute budding —

    (e)   Chip budding —

Asexual methods of plant propagation :

It is also called as vegetative methods, in which the vegetative parts of plants such as leaves, stems, roots are made use of for propagating in new individual plants. There are two main types :

(A) Common origin methods — 3 methods

(B) Separate origin methods — 2 methods

{A} **Common origin methods :**

[I] **Plant parts generally detached before rooting 2**

   (a) Separation -2 subtypes

   (b) Division -5 subtypes

(a) **Separation (Bulbs and Corms)**

   (i) **Bulbs :** It is modified underground stems consisting of basal disc from the base of which the roots are formed.

   e.g., Scally bulbs — Onion, lily

   Tunicated bulbs — Garlic, tuberose,

   (ii) **Corms :** It is specialized underground swollen base of stem resembling like bulbs. Morphologically it consist of nodes and internodes. Roots came out from basal concave surface and shoots from dorsal surface, e.g., suran, gladioli.

(b) **Division** – Rhizomes, offsets, tubers, suckers & runners.

   (i) **Rhizomes :** It is a specialized structure in which the main axis of plant grows horizontally just below or on the surface of ground. 1t is fleshy, cylindrical in shape and bears nodes, internodes, auxiliary buds which gives rise to shoots and roots, e.g., Turmeric, ginger, canna.

   (ii) **Offsets :** It is lateral shoot or branch which develops form the base of main stem and bears nodes, internodes, auxiliary buds, e.g., Cardamum, pine apple.

   (iii) **Tubers :** It is thickened, modified underground stem or root. 1t has eye buds in the axils of scale like leaves, e.g., stem tubers – potato

   Root tubers – sweet potato, dahalia.

   (iv) **Suckers :** It is specialized underground structure which has pseudostem and which grows from underground stem of mother plant vertically above the ground, e.g., Banana.

   (v) **Runners :** Specialized stem which develops from auxiliary bud of plant. 1t grows horizontally along the ground and forms a crown of roots at the node which comes in contact with the soil. This crown throws out shoots, roots, e.g., Hariyali, strawberry.

[II] **Plant parts invariably detached before rooting (cuttage) – stem, root, leaf cuttings.**

(a) **Stem cutting :** Branch is first detached from he mother plant and then induced to produce roots.

   (I) Hard wood cuttings – Grape, rose, bouganvellia

   (II) Semi hard wood cuttings – Croton, arelia, panax

   (III) Softwood cuttings – Coleus, geranium, pilea

(b) **Leaf cuttings :** Certain ornamental plants and monocots, dicots are propagated easily with this method.

   (I)   leaf cutting by using petiole – pepromia

   (II)   leaf cutting by using marginal buds – bryophyllum

   (III)   leaf cutting by notching & main veins – begonia

(c) **Root cuttings :** Roots are used, e.g., bread fruits.

[III] Plant parts invariably not detached before rooting (Layerage)

It is the vegetative method of plant propagation where in the vegetative parts like stem (branches) of the plant is first induced or forced to produce roots while still attached to the mother plant. The branch which roots out while still attached to the mother plant is called as "layer" and has the same genetic constitution as that of mother plant.

Methods of Layerage

(a) Air layer or morcottage or gootee.

(b) Air layer with plastic

(c) Tongue layer

(d) Tip layer

(e) Compound layer

(f) Mound layer.

(B) Separate origin methods : The single individual plant is developed by use of the plant parts from two separate mother plants. There methods are grafting and budding.

(a) Grafting : It is and art of incertion of scion into the stem of the rootstock in such a way that union takes place and the combination continuous to grow. The scion used is taken from the desired tree which is called as scion tree or mother plant. The rootstock is raised separately for this purpose. Rootstock and scion combination are the two components of a graftage.

Scion : Is the short piece of detached shoot containing several dormanent buds, which when united with the stock, comprises the upper portion of the graft and from which will grow the stem and branches of the new grafted plant. 1t should be the desired variety and free from diseases.

Stock (Rootstock) : Is the lower portion of the graft which develops into the root system of grafted plant. 1t may be a seedling, rooted cutting or a layer plant.

Methods of grafting

(A) Scion attached methods : The scion is kept attached to the mother plant till the graft union takes place and then graft is separated in stages taking cuts on scion below the graft union and on the rootstock above the graft union. This principle is followed in following methods.

(i)   Simple approach or Inarching

(ii) Saddle grafting

(iii) Tongue grafting

(B) Scion detached methods : The scion is first detached from mother plant and then inserted into the rootstock so as he union takes place and combination continuous to grow. These methods are:

(I)   Veneer grafting

(II)   Saddle grafting

(III) Wedge grafting

(IV)   Whip grafting

(V)   Tongue grafting

(VI)   Stone grafting

(C) Methods of grafting on established trees : These are the methods which can be successfully adopted to convert the inferior crown of the established plants into superior crown. These are:

(i)   Side grafting

(ii) crown grafting

(iii) Top working–

 (a)   by inarching

 (b)   by crown grafting

 (c)   by forkert budding.

(D) Renovation methods of grafting : These are the grafting methods which are adopted to rejunvate the old trees.

(i) Bridge grafting

(ii) Butteress grafting

(b) Budding : It is an art of insertion of a single mature bud into the stem of the root stock in such way that union takes place and combination continuous to grow. It is grafting of a single individual bud instead of whole bud stick or scion as is done in case of grafting. There are several techniques or methods of incretion of bud into the rootstock. The adoption of any of the methods depends upon the plants to be budded, situation, facilities and sources available.

Methods of budding

  (i)   Shield budding – 'T', & I budding

 (ii)   Patch budding

(iii)   Flute budding

(iv)   Ring budding

 (v)   Forkert budding

10. Define Horticulture? What are the branches of horticulture?

The word 'Hortus' is a latin word meaning 'Garden' and culture means cultivation, Thus cultivation of garden crops is included in Horticulture. There are four branches of Horticulture.

  (i)   Pomology (Fruit Science)

 (ii)   Olericulture (Vegetable Science)

(iii)   Preservation of fruits & Vegetables

(iv)   Ornamental gardening, i.e., Floriculture & Landscape gardening

Pomology : It includes the study of different fruits crops like mango, banana, grapes, citrus, etc. All aspects regarding the cultivation of fruit crops such as planting, training, pruning, control of pest and diseases, etc in addition to this there is selection of site, selection of fruit crops & planting methods.

Olericulture : It includes the study of different vegetable crops. All aspects from selection of vegetable till the marketing of products are studied.

Preservation of fruits & vegetable : This is the industrial part of the subject in which the principles and methods of different products of fruits and vegetables are studied.

Ornamental gardening : It include two parts a&b.

(a)   **Floriculture :** It is art of growing, arranging and selling of flowers. Designing the gardens and arrangement of cut flowers is also included.

(b)   **Landscape gardening :** This consist of planning and arrangement of homes, farm sheds, public areas, business establishment, play grounds. 1t includes placement of buildings, walks, drives, fences, recreation areas. The advantages of existing natural things and different structures is many times utilized as a part of landscape garden. Also it is beautification of natural areas without disturbing the nature.

11. Which are the important fruit zones of Maharashtra

Zone	Rainfall	Districts	Fruits crops
1. Warm coastal region	80-200"	Ratnagiri, Raigad, Thane, Sindudurg.	Mango, cushownut sapota, banana, pineapple, jackfruit, Arceanut, coconut, spice crops.
2. Western Deccan region	30-50"	Parts of Kolhapur, Nasik, Satara, Sangali, pune.	Mango, banana, citrus, sapota, guava, jackfruit
3. Eastern Deccan region	15-25"	Ahmednager, Jalgaon, Dhule, Solapur, A' bad, Bhir, U' bad, East Pune	Mosambi, limes, papaya, banana, pomegranate. Fig.
4. Western Vidharbha and marathwada.	20-30"	Buldhana, Akola, Yeotmal, Parbhani, Nanded, A' bad.	Mango, banana, Grape, mosanbi, Santra, Fig, guava, custard apple, Easterm phalsa.
5. North east regien of vidharbha	30-40"	Amravati, wardha, Nagapur, Chardrapur, Akola, Bhandara.	Santra, mosambi, mango, papaya, guava, limes.

12. What is unfruitfulness? What are the factors associated with the unfruitfulness.

It is phenomenon in which even though, tree produces blossoms, fails to develop fruits to maturity and to produce satisfactory crop. Or It is an inability of producing mature fruits such tree is called unfruitful.

Factors associated with unfruitfulness.

(A) INTERNAL FACTORS

[I]   Evolutionary tendencies

(i)   1mperfect flowers – monoecious & diocious flowers

(ii)   Heterostyly – difference in height of style & filament

(iii)   Dichogamy – maturity of pollens at different times

(iv)   Structural pecularities – closed flowers

(v)   Abortion of pistil or ovule

(vi)   Impotance of pollens – non-viable pollens

[II]   Genetic influence

(i)   sterility and unfruitfulness due to hybridity

(ii)   Incompatibility of pollens and ovules.

[III] Physiological influence

(i)   Unfruitfulness due to slow growth of pollen tube

(ii) Due to early or late pollination

(iii) Nutritive conditions within the plant body.

(B) EXTERNAL FACTORS

(i)   Nutrient Supply vii. Light

(ii)   Pruning and grafting viii. Irrigations

(iii) locality ix. Rains at blossoming

(iv)   season x. winds

(v)   Age & vigour of plant xi. Insects & diseases

(vi)   Temperature xii. Spraying during blossoming.

13. What is C:N ratio? Explain its proportion in fruiting of fruits trees

The ratio of accumulation of carbohydrates and nitrogen in the plant body plays an important role in changing the bud to be fruitful. For the flower bud initiation a certain ratio of carbohydrates and nitrogen is required to exist.

(a) C:N both are equal in proportion :

Carbohydrates and nitrogen are more and in almost equal proportions i.e. there is moderate rate of accumulation and hence there is normal fruiting.

(b) More of carbohydrates and less of nitrogen :

Then there is small growth with early but less fruiting. Accumulation of carbohydrates is higher than its utilization.

(c) More nitrogen and Less carbohydrates :

Then the plant is very vigorous with less and late fruiting. The rate of utilization of carbohydrates is more than its accumulation.

(d) Both Carbohydrates and Nitrogen are low, plant remains stunted with absolutely no flowering.

14. What is Pruning in Grapes? What are its purposes?

Pruning is an art and science of removing sufficiently certain part of plant with a view to divert sap flow towards fruiting area to induce vegetative and fruitful growth.

Grape vines are pruned for three purposes

1. Young vines is pruned for training where side shoots removed to facilitate development of single stem. The growing point of vine is pinched at desired height to develop primary and secondary arms.

2. Pruning for vegetative growth – <u>April pruning</u> – or foundation pruning, which is done after harvest of crop, only one bud is maintained as a foundation for new growth. This is done in 15 March to 15 April.

3. Pruning for fruits or <u>October pruning</u> – this is done in 130-160 days after April pruning i.e. between 15 Sep to 15 Oct. Matured canes are pruned keeping definite numbers of buds according to vigour of variety and thickness of cane.

In North India, because of severe winter, vine remains dormant during winter and put forth new growth only during spring and therefore pruned only once before spring. Under southern conditions vine grows continuously because of warm temperature throughout the year therefore pruned twice a years.

15. Which are the Systems of planting of fruit crop orchards?

Layout means locating the position of trees, roads and buildings in the orchard being established, and systems of planting refer to the orderly ways of planting the trees. It is desirable to have the trees planted in a systematic way because.

—— Interculturing and irrigations becomes easy

—— Equal distribution of area under each crop

—— Least wastage of land

—— Supervision easy & effective

—— There is room for systematic extension of the orchard.

There are Five Systems of Plantings

1. Rectangular system

2. Square system – all over India it is adopted

3. Diagonal system or Quinounx

4. Hexagonal System

5. Contour system

   (1) **Rectangular system :** In this method trees are planted in a straight line (row) giving a straight angle only. It is suitable for grapes in Maharashtra.

(2) **Square system :** In this method a tree is planted on each corner of square. Intercultivation in both way is easy in this method. Plant to plant and row to row distance is same.

(3) **Diagonal system :** In this method at the centre of each square a tree is planted (filler tree)

(4) **Hexagonal system :** Also known as triangular system In this system the trees are planted in each corner of an equilateral triangle system ways, the six trees form a hexagon with he seventh tree in the centre. This system allows planting of 15% more trees than square system.

(5) **Contour system :** It is usually followed on the hills with high slopes. In this method trees are planted across the slopes. Contour trenches are dug out across he slopes and trees are planted. The points having same altitude are connected together by a line & the trees are given spacing on this line. This retards he ill effects of erosion.

16. Define Plant Growth regulators? Write classes and uses of plant growth regulators?

Plant growth regulators are the organic chemical compounds which modify or regulate physiological processes in an appropriate measure in the plants when used in small concentrations. They are readily absorbed and more rapidly through the tissues when applied to different parts of the plant. These chemicals are specific in their action.

Classes of growth regulators

(1) Auxins – IAA, IBA, NAA, 2-4-D.

(2) Gibberellins – GA1 to 74

(3) Cytokinins – Kinetin, zeatin

(4) Abscisic acid – ABA, dormin

(5) Ethylene – Ethrel, Ethephon.

Uses of plant growth regulators

• Propagation of plants through harmone treatment to cuttings and scions.

• Prevention of pre-harvest fruit drops

• Increasing fruit set and parthenocarpy

• Inhibition of buds to prevent sprouting.

• Prolonging dormancy to prevent damage due to frost.

- Control of flowering
- Defoliation of plants.
- Prevention of leaf fall.
- Thinning of fruits.
- Selective weed killers.
- Prolonging shelf life of fruits.
- Ripening of fruits.

17. Define vegetables & give its importance & scope

Vegetables – any herbaceous plant or their plant parts which is being consumed in fresh condition. The edible portion may be root, tuber, bulb, stem, petiole, leaf, flower, or flower buds, partially developed seed receptacle, mature seed or fruit. The use of vegetable varies according to the kind or vegetable.

Economic Importance of Vegetables

- They are rich sources of protective elements, minerals, vitamins &other chemical substances.
- Per unit acre yield of vegetables is very high.
- Vegetables are an important source of farm income.
- They have high aesthetic value.
- More vegetables can be grown in a year.
- Vegetables fulfils the daily requirement.
- Maintain fertility of soil.
- Provides fodder to animals.
- Grown as a rainfed crops.
- Have medicinal properties.

Scope of Vegetables

- To meet the daily requirement of vegetables in diet.
- Because of development of new cities.
- Advanced cultural practices.
- Quick transport , irrigation, storage facilities.
- Companion cropping – same cultural operations, e.g., cole crops.

- Export potential
- Vegetable seed production
- Vegetable processing.
- Off season cultivation
- Shadnet cultivation

18. Write the Classification of Ornamental plants?

The lifecycle initiated from seeds and completed with the formation of mature nature seed. The life cycle does not end with the production of seed but plants still remain to live for successive number of years. On the basis of their life cycle plants are classified into three majer groups.

(1) **Annuals :** Plants which completes their life span from seed to death within a single season, e.g., Rainy, summer winter season flowers, vegetables.

(2) **Biennials :** Plants which complete their life span from seed to death in two seasons, e.g., Onion, carrot, cabbage, daisy, beets.

(3) **Perenninals :** Plants which complete their life span from seed to death in three or more seasons. These are classified in two groups.

   I. Herbaceous – Plants with non-woody stem.

   (a) Bulbs, corms, tubers rhizomes, cactus

   (b) Gerberas, geraniums

   (c) Orchids, ornamental foliage plants

   II. Woody plants – Two groups

   (a) Evergreen – retains their leaves e.g. shrubs, trees creepers.

   (b) Deciduous – which sheds their leaves e.g., trees.

19. What is Plantation Crops?

The crops which have industrial value and requires processing before consumption are known as plantation crops. These crops requires more man power and needs care and maintenance throughout the year. India has a well reputation as land of plantation crops like coconut, coffee, arecanut, tea and spices. India is a major exporter of plantation products in the world. These crops has greeter importance in social life of man, e.g., coconut, arecanut, turmeric are used in social functions such or puja, wedding, inaugural functions and different ceremonies.

20. Which are the selection criteria for selection of mother palm in coconut?

- It should be located in centre of the garden.
- Palm should have stout, straight trunk with closely spaced leaf scares, 30-40 average sized fronds, well oriented on crown and large no of inflorescences with short stalks distributed entirely round the crown.
- Palm should have good distribution of nuts, higher weight of husked nuts, higher percentage of fruit set, large no of spikelet's.
- Palm should be regular bearer.
- Age of palm should be 10-25 years.

21. Which are the criteria for selection of seedlings in coconut?

It has been proved that seedling selection alone can increase 10% yield. Following characters should be considered while selection of seedlings.

- Early germination of nuts.
- Broad and dark green leaves.
- Early splitting of leaflets.
- Short and broad leaf stalks.
- Straight and short stem with good girth at collar.
- Large number of roots.
- Age of seedling should be 9-12 months.
- It has 6-7 leaves.

22. Which are the criteria for selection of seednut & mother palm in Arecanut?

- Selection of mother palm should be done only from reputed gardens having high yielding and regular bearer trees (20-40 years).
- Early bearing of mother palm is related to higher % of fruit set.
- Poor yielder mother trees should be rejected.
- Fully ripe nuts from middle bunch should only be selected and lowered down by means of rope.

23. Enlist the important garden features?

- Garden walls
- Lawn

- Fencing
- Steps
- Drives and paths
- Hedges
- Edges
- Arches
- Pergolas and Bowers
- Terrace and terrace gardening
- Paved garden
- Dry walls – walls with different plants
- Carpet bedding
- Flower beds
- Shrubbery – growing shrubs
- Borders
- Rockery or rock garden
- Water garden
- Marsh or bog garden
- Sunken garden
- Gardening in the shade
- Roof gardening
- Green house, conservatory and lath house
- Garden adornments – attractive features, e.g., fountains, ponds, pools.
- Topiary work
- Bonsai – dwarf potted trees.
- Indoor plants.

24. Define garden. What are the types of gardens?

Garden can be defined as an area established with plant valuables and pleasurable around or adjacent to house or building.

A garden is work of art. It is skilled arrangement and design of plant over the area making a design or pattern or picture as it was that forms the garden.

Garden offers light, pleasant recreation after the days hard work and business tension and is hence necessity of modern life.

Garden Types

1. Landscape gardens
2. Formal gardens
3. Informal gardens
4. Wild gardens
5. Public parks
6. Vegetable gardens
7. Dish garden
8. Desert garden
9. School garden
10. Dryland garden

(1) **Landscape gardens :** It is the beatification of natural areas without disturbing he original things. The art of designing the landscape is known as landscape gardening.

(2) **Formal gardens :** It is laid out in a symmetrical or geometrical pattern. The roads, paths, flower beds, borders and shrubbing are arranged in a geometrically designed bed.

(3) **Informal gardens :** In this the design features are arranged in a natural way without following any rules. The plan is laid out as per well thoughts.

(4) **Wild garden :** It is comparatively recent type of garden expanded by Willam Robinson in the last decade of the 19th century. The concept of wild garden is not only against the all formalism but it also breaks the rule of landscaping. The main idea was to naturalise the plants in shrubberies.

(5) **Public parks :** It may be informal or landscaping including wild animals (Zoo), fishes, small railways and other ornamentals.

(6) **Vegetable gardens :** Sometimes vegetable gardens also gives good ornamented show, Its types used for ornamental purpose are kitchen and floating gardens.

(7) **Dish gardens or plantones :** These are created by placing plants together so closely that they have no room for further growth. Little drainage space is provided in he container. It contains tropical plants, cacti, succulents.

(8) **Desert gardens :** Planters constructed with air tolerant plants such as cacti and desert succulents are grown in fast draining medium.

(9) **School gardens :** Small garden grown by students near the school is also a type of garden. It adds skill to them.

(10) **Dryland gardens :** Garden plants are subjected to water stress condition. Stress or drought tolerant plants are cultivated.

25. Information of fruit crops

Sr. No.	Name	Propagation method	Spacing	No. of plants /ha	Fertilizer dose NPK for full grown tree	Yield / plant	Important varieties
1.	Mango	Stone grafting, Inarch grafting, softwood grafting	Light soil, 9x9 m heavy soil, 10x10 m	121 100	1.0 : 0.5 : 1.0 kg/tree	500 - 1000 Fruit/tree	Alphanso, Keser, Pairi, Totapuri, Vanraj, Ratna
2.	Guava	Layering (tongue)	8 x 8 m	156	900:300:300 g/plant	700-1500 Fruit/tree	Sardar, Allahabad Safeda, Chittidar, Arka mridula
3.	Pomegranate	Layering (gooty)	4.5 x 3 m	740	600:250:250 g/plant	100-150 Fruit/tree	Mrudula, Phule Arkta, Bhagwa, Ganesh, Ruby, Joyti
4.	Sweet orange	Grafting on Rangpur lime or marmalade rootstock.	6 x 6 m	277	800:300:600 g/plant	800-1000 Fruit/tree	Nuceller, Phule mosambi
5.	Kagzi lime	Seedlings	5 x 5 m 6 x 6 m	400 277	800:300:200 g/plant	1500-2000 Fruit/tree	Sai-Sarbati, Phule Sarbati, Vikram, Prumalini
6.	Banana	Tissue cultured seedling or Pseudosuckers	150 x 150 cm 135 x 135 cm 150 x 135 cm	4444 5400 4866	100:40:100 g/plant	15-20 kg/tree or 50-60 t/ha	Basari, Harisal, Lal velchi, Shreemanti, Grand naine
7.	Grapes	Cuttings or Grafting on different root stocks	3 x 3 m 3 x 1.5 m	1111 2222	900:500:700 kg NPK per ha.	20-40 ton/ha	Thompson seedless, Tas-A-Ganesh, Sonaka, Sharad seedless, Flame seedless.
8.	Sapota	Grafting on khirni rootstock	10 x 10 m	100	1.0:0.5:0.5 kg/plant	1500-3000 fruit/plant	Kalipatti, Cricket ball, pillipatti.
9.	Fig	Cutting or layerings (gooty)	5 x 5 m	400	900:250:275 g/plant	15-20 kg per plant	Poona fig, Dinkar, Conadria, Excel
10.	Papaya	Seedlings	2.25 x 2.25 m	2000	200:200:200 g/plant	30-50 fruit per plant	Washington, Co-2, Pusa delicious, Solo.
11.	Coconut	Seedlings	10 x 10 m	100	1.0:0.5:1.0 kg/plant	80-100 fruit / plant	Banawali, Pratap, TXD
12.	Ber	Budding	6 x 6 m	277	250:200:50 g/plant	75-100 kg/plant	Umran, Chuhara, Gola, Kadaka, Mehrun

Contd...

Contd...

Sr. No.	Name	Propagation method	Spacing	No. of plants /ha	Fertilizer dose NPK for full grown tree	Yield / plant	Important varieties
13.	Custard apple	Seedlings, Grafting or budding	4 x 4 m5 x 5 m	625400	250:125:125 g/plant	50-100 fruit/plant	Balanagar
14.	Aonla	Patch budding	8 x 8 m	156	1375:1600:400 g/plant	100-150 kg/plant	Kanchan, Neelam, Krishna, NA-7
15.	Jamun	Seedlings or Patch budding	10 x 10 m	100	500:250:250 g/plant	100-150 kg/plant	Konkan Bahadoli
16.	Tamrind	Seedlings or inarch grafting	10 x 10 m	100	500:250:250 g/plant	50-150 kg / plant	Pratisthan, N0263

Information of vegetable crops

Sr. No.	Name	Seed rate kg/ha	Sowing time and season	Spacing	Fertilizer dose NPK kg/ha.	Duration (days)	Yield q/ha	Important varieties
1.	Onion	8-10	*Kharif*-June-July Rangda-Sept.-Oct. *Rabi*-Nov.-Dec. Summer-January	for *kharif* 15 × 10 cm for Rabi 12.5 × 7.5cm	100:50:50	110-120	*Kharif*-15-20 t/ha *Rabi* and Rangada 20-25 t/ha. Summer 30-35 t/ha.	For *kharif* N-53, B-780, Phule Samarth, Phule Safed, AFDRFor rabi-N-2-4-1, AFLR, Arka niketan
2.	Chilli	1.00	*Kharif*-June-July Summer-Jan-Feb	60 × 45 cm	100:50:50	180-200	Green-200 q/ha dry-25 q/ha	Phule Joyti, Pant C1, Phule Mukta, Phule Sai, Pusa Jwala.
3.	Tomato	0.400	*Kharif*-June-July *Rabi*-Sept.-Oct.	90 × 30 cm	200:100:100 for varieties 300:150:150 for hybrids	150-160	30-40 ton/ha 60-70 ton for hybrid	Dhanshree, PED, Phule raja, Bhagayashree, Pusa rubi, Marglobe
4.	Brinjal	0.500	*Kharif*-June-July Summer-Dec-Jan.	90 × 75 cm 90 × 90 cm 75 × 60 cm	100:50:50	180-200	25-30 ton/ha	Krishna (F_1), Manjari gota, Phule Harit, Pragati.
5.	Cabbage	0.600	Sept-Oct	45 × 30 cm	160:80:80	65-80	25-30 ton/ha	Goldaen Acre, Pride of India, Pusa drum bead.
6.	Cauliflower	0.600	May-December	60 × 45 cm	150:75:75	70-100	20-25 ton/ha	Pusa Katki, Kunwari, Pusa synthetic, snowball 16 , Pusa Dipali
7.	Potato	800-1500 kg	*Khrif*-June-July *Rabi*-Oct-Nov.	45 × 30 cm	150:60:120	90-110	20-30 ton/ha	Kufri chandramukhi, Kufri Jyoti, Kufri lavkur, Kufri sinduri
8.	Okra	12-15 kg	*Kharif*-June-July Summer-Jan-Feb.	30 × 20 cm	100:50:50 30 X 15 cm	110-125	12-15 ton/ha	Parbhani Kranti, Arka anamika, Phule Utkarsha Akola Bahar, Varsha Uphar
9.	Pea	80-90 kg	Sept.-Oct.	30 × 15 cm	20:60:60	80-90	4-6 ton Green pods/ha1-2 ton dried peas/ha	Boneville, Arkel, VL-3, VL-8, PH-1

Contd...

Contd...

Sr. No.	Name	Seed rate kg/ha	Sowing time and season	Spacing	Fertilizer dose NPK kg/ha.	Duration (days)	Yield q/ha.	Important varieties
10.	Radish	8-10 kg	Sept.-Nov.	30 × 15 cm	30:20:80	45-60	15-20 ton/ha.	Japaness white, white long, Pusa deshi, Pusa chetaki, Pusa Reshmi, Ganesh Synthetic
11.	Methi	35-40 kg	Sept.-March	Broadcasting	80:40:40	30-40	10-12 ton/ha.	Pusa early bunching, Kasturi, RMT-1, Methi No. 47.
12.	Palak	8-10 kg	Sept.-Dec.	Broadcasting	40:40:40	90-115	15-20 ton/ha.	Pusa Joyti, All green
13.	Biter gourd	2-2.5 kg	Kharif-June-July Summer-Jan-Feb.	1.5 × 1.0m	100:50:50	180-200	20-25 ton/ha	Coimbature long white, Hirkani, Phule green gold.
14.	Ridge gourd	2-2.5 kg	Kharif-June-July Summer-Jan-Feb.	1.5 × 1.0 m	100:50:50	140-150	15-20 ton/ha	Pusa nasdar, Konkan Harita, Phule Sucheta
15.	Water melon	2.5-3 kg	Summer-Jan.-Feb.	2 × 0.5 m	100:50:50	90-120	40-50 ton/ha	Sugar baby, Arka Joyti, Arka manik
16.	Muskmelon	1.5-2 kg	Summer-Jan.-Feb.	1.5 × 0.5 m	100:50:50	80-100	20-25 ton/ha	Pusa Sarbati, Harama dhu, Punjab Sunhari, Punjab Rasila, Durya pura sel.
17.	Cucumber	1.1.5 kg	Kharif-June-July Summer-Jan-Feb.	1.0 × 0.5 m	100:50:50	90-120	15-20 ton/ha	Poona Khira, Poinesette, Sheetel, Himangi, Phule Suthangi
18.	Bottle gourd	2.2.5 kg	Kharif-June-July Summer-Jan.-Feb.	3.0 × 1.0 m	100:50:50	180-200	40-50 ton/ha	Samarth, Pusa naveen, PSPL, PSPR, Varad.
19.	Cluster bean	14-24 kg	Kharif-June-July Summer-Jan-Feb.	30 × 15 cm or 45 × 15 m	35:60:60	90-110	5-6 ton/ha	Pusa Sadabahar, Pusa Mosami, Pusa Navbahar
20.	Fresh bean	40 kg	Kharif-June-July Summer-Jan.-Feb.	45 × 30 cm 60 × 30 cm	50:110:110	90-110	1-1.5 ton seed yield	Contender, Pusa Parvati, Arka Komal, Pant Anupama, Phule Suyash, Phule Surekha.

Contd...

Contd...

Sr. Name No.	Seed rate kg/ha	Sowing time and season	Spacing	Fertilizer dose NPK kg/ha.	Duration (days)	Yield q/ha.	Important varieties
21. Garlic	6 quintal /ha	*Rabi*-Oct.-Nov	10 × 10 cm	100:50:50	130-150	9-10 ton/ha	Godawari, Sweta, Yamuna Safed, Agri-found white Phule Baswant,
22. Dolichus bean (wal)	2.5 kg for pole type 20-30 kg for bush type	*Kharif*-June-July *Rabi*-Sept.-Oct.	2.0 × 1.0 m for pole type 60 × 30 cm for bush type	60:60:60	180-200 for pole type, 100-120 for bush type	20-25 ton/ha, pole type, 8-10 ton/ha for bush type	Pole type-Phule Gouri, Phule Ashwani. Dasara, Deepali:Bush type-Konkan Bhushan, Arka Jay, Arka Vijay
23. Sanke gourd, (Padval)	2.5 kg / ha	*Kharif*-June-July Summer-Jan-Feb.	2.5 × 1.0 m	100:50:50	180-200	25-30 ton/ha	Phule Vaibhav, Konkan Sweta
24. Drumstick	Seedlings or cuttings	*Kharif*-June-July	5 × 5 m	75:50:50 g NPK/plant	Perennial	25-50 kg pods/tree	CO 1, CO 2, PKM-1, Konkan-Ruchira
25. Ginger	15-20 quintal sets	April-May	20 × 20 cm 45 × 20 cm	75:30:30 kg NPKV/ha	9-10 months	150 quintal	Mahim local

Information of flower crops

Sr. No.	Name	Propagation method	Spacing	Seed rate / No. of plants/ha	Fertilizer dose	Yield q/ha.	Important varieties
1.	Rose	Budding (Rosa indica)	75 x 75 cm 1.5 x 0.45 m	17,770 14,800	60:40:40 g NPK/plant	3 to 4 lakhs flowers	Gladiator, Superstar, Piece, Rootstock) Happiness, Sonia, Princes, Ifeltower, Pari porch.
2.	Chrysanthemum	Suckers	30 x 30 cm	1,11,111	100:50:30 kg NPK/ha	6-10 thousand kg flowers	Raja, Pandhari rewadi, Supreme, Perrete
3.	Aster	Seed	45 x 30 cm	1 to 1.5 kg seed/ha	250:100:100 kg NPK/ha	2-3 lakh flowers,bundles	Alandi mix, queen of market, Astich plum, Phule Aster 6,7,8,9
4.	Nishigandh (Tuberose)	Mother sets	30 x 20 cm	1-1.5 lakh sets	200:300:300 kg NPK/ha	1.5-3.0 lakh flowers	Single, double, Phule Rajani
5.	Gallardia	Seed	30 x 30 cm	200 g seed/ha	60:40:25 Kg	2 to 3 NPK/ha	lakh bundles Indian Chief, Dazlar, Sangini
6.	Gladiolus	Sets, corms	30 x 10 cm	1.5-3.0 lakh sets	250:200:200 Kg NPK/ha	1.5 to 3.0 lakh flower stalks	Suchitra, Supra, Nazrana, Sancerre, Yellowstone, Phule Ganesh, Phule prerena, Pule Tejes, Phule Neelrekha

Information of spice crops

Sr. No. Name	Propagation method	Spacing	Seed rate / No. of plants/ha	Fertilizer dose	Yield q/ha.	Important varieties
1. Black papper	Cuttings/vines	3 x 3 mt standards viz. coconut and Arecanut intercrop	1111	10 kg FYM 100:40:140g NPK per vine per year	2-3 kg per vine per year	Cherikan iakadam, Uthiranktta, Panniyur-1, Balanketta
2. Cinnamon	Seedlings/ cutting	2 x 2 mt inter crop in coconut garden	2500	20 kg FYM 150:75:150g NPK / plant	200-300 kg bark and 2-3 kg leaf oil/hectare/year	Cinnamon zeylanicum Cinnamon cassia Cinnamon burmanii Cinnamon campura
3. Clove	Seedlings	In arecanut and Coconut gardens 2.7×2.7 mt	1371	20 kg FYM 150:75:150g NPK /plant/year	2 kg per plant per year	Penang cloveZanzibar clove
4. Turmeric	Rhizomes (Mother sets,	45 x 20 cm 37.5 to 30 cm	Seed rate 2500 kg/ha sets	25 ton FYM 120:50:50NPK kg/ha	250-300 q/ha fresh rhizomes 60-70 q/ha dry rhizomes	Rajapuri, Salem, Waigaon, Tekurpeta, Erode, Phule Swarupa, Alleppy.

Information of plantation crops

Sr. No.	Name	Propagation method	Spacing	Seed rate / No. of plants/ha	Fertilizer dose	Yield q/ha.	Important varieties
1.	Rose	Budding (Rosa indica)	75 x 75 cm 1.5 x 0.45 m	17,770 14,800	60:40:40 g NPK/plant	3 to 4 lakhs flowers	Gladiator, Superstar, Piece, Rootstock) Happiness, Sonia, Princes, Ifeltower, Pari porch.
1.	Coconut	Seedlings	10 x 10 mt 8 x8 mt	100 156	50 kg FYM 1.0 : 0.5:1.0 kg NPK/ plant per year for adult palm.	60-80 nuts/palm /year	WCT, Laccadive Oridinary, Kappedam, Spicata, Banwali, Dwarf orange, TXD, DXT
2.	Arecanut	Seedlings	3 x 3 mt	1111	12 kg FYM 150:75:150 g NPK/palm/year adult palm	14 kg nuts/palm /year	Mangala, Shriwardhani, Andaman, VTL-11, VTL-17, Mohitnagar
3.	Cacao	Seedling/ cuttings	3 x 3 mt & intercrop in coconut and arecanut gardens	1111 filler tree 7-9 mt spacing	30 kg FYM 100:80:140 g NPK/tree/year	0.5 to 1.5 kg beans per plant per year	Criollo, Trinitario Forastero
4.	Coffee	Seedlings / Cuttings	3 x 3 mt	1111	348:260:348 NPK kg/acre	450-550 kg/ha	Coffee ArabicaCoffee RobustaCoffee Libarica
5.	Ginger	Rhizomes	20 x 20 cm 30 x 30 cm	Seed rate 1000-1200 kg sets/ha	25 ton FYM 75:30:30 NPK kg/ha	5-6 tons/ha Green ginger	Rode-Janeriro, Maran, Nadia, Tinladium, Himachal Pradesh, Mahin local.

REFERENCES

1. Neeraj Pratap Singh, 2005. *Basic concepts of fruit science.* pp. 1-667. Published by International Book Distributing Co., Charbagh, Lucknow, U.P. India. (ISBN 81-8189-059-0).

2. Amita Bishnoi and Milind Singh, 2005. *Agricultural Dictionary of Scientific and Technical Abbreviations.* pp.- 1-256. Published by Kalyani publishers, New Delhi. (ISBN 81-272-1937-1).

3. Chadha, K.L. 2002. *Handbook of Horticulture* Published by ICAR, New Delhi. pp. 1-993.

4. Thamburaj, S. and N. Singh, 2001. *Textbook of Vegetables*, Tuber Crops and Spices. pp. 1-468. Published by ICAR, New Delhi.

5. Bose, T.K. and S.K. Mitra, 1996. *Fruits – Tropical and Subtropical.* pp.1- 803. Published by Naya Prakash, Calcutta.

6. Adams, C.R. and M.P. Early, 2004. *Principals of Horticulture.* pp. 1-220. Published by Elsevier India Pvt. Ltd., New Delhi, (ISBN-10 : 81-312 (00817).

7. Hartmann, H.T. and D.E. Kester 1972, *Plant. Prorogation – principles and practices.* PP. 1-702 published by Prentice-Hall of India private Limited, M-97, Connaught Circus, New Delhi – 1.

8. Edmond, J.B., Musses, A.M. And Andews, F. S. (1957). *Fundamentals of Horticulture.* Mc Graw Hill Book Co., New York.

9. Denison, E.T. (1958). *Principles of Horticulture.* McMillian Publishing Co., New York.

10. Bose, T.K., Mitra, S.K., and Sanyal, D. (2002). *Plant Propogation Principles and Practices.* Naya Udyog, Calcutta.

PART B

KEY NOTES
ON
AGRICULTURE EXTENSION

1

TERMINOLOGY

Term	Definition/Explanation
Adopter	Adopter is a person who has continued full use of innovation.
Adoption	Is a decision to make full use of an innovation as the best course of action available?
Agricultural Extension	Is not only imparting knowledge & Securing adoption of a particular improved practice but also aims at changing the outlook of the farmers to the point where they will be receptive to & on their own initiative, continuously seek means of improving their farm, occupation, home & family life in totality. (National Commission on Agricultural).
Agricultural Extension Assistance	Agricultural Extension Assistance to farmers to help them to identify and analyze their production problems and to become aware of the opportunities for improvement.
Agricultural Information Centre Aim	Agricultural Information Centers are set up at district head quarter. It collects information suitable to the farmers. Aim is broad objective.
Andragogy	It means art and science of helping adults to learn.
Association	It is a group of persons organizing themselves for the purpose of fulfilling certain interests.
Attitude	It is a mental state of readiness, organized through experience, exerting a directive & dynamic influence upon the individual's response to all objects & situations with which it is related.
Audience	Audience is the intended receiver of messages.
Audio-Visual Aids	The instructional devices through which the message can be heard & seen simultaneously are known as Audio-visual aidse, e.g., Movie picture.

Term	Definition/Explanation
Audio Aids	The instructional devices through which the message can only be heard are known as audio aids, e.g., Radio.
Audio visual Aid	Audio visual aid is an instructional device that can be heard and seen.
Behaviour	Of an individual's refers to anything an individual does.
Beliefs	Which are closely related to values, are the mental convictions one has about the truth or actuality of something (Soprano).
Bulletin	It is a printed, bound booklet with a number of pages, containing comprehensive information about a topic.
Calendar of Work	It is a plan of work arranged chronologically according to the time, when each step of work is to be done.
Campaign	Is an intense educational activity for motivating & mobilizing a community to action, to solve problem or satisfy a need urgently felt by it.
Cast system	Collection of family being common value which usually denotes association with specific occupation and life.
Caste	When class is somewhat hereditary it is called as caste.
Channel	It is the medium through which information flows from a sender to one or more receivers, e.g., Newspaper, Radio, TV. Channel is the physical bridge between the sender and the receiver of a message.
Charge Agent	A professional person who attempts to influence adoption decisions in a direction, he feels desirable.
Communication	Is the process by which two or more people exchange ideas, facts, feelings or impressions in ways that each gains a common understanding of the meaning, intent, & use of messages. Communication is process by which two or more people exchange their ideas, facts, feelings or impressions in ways that each gains a common understanding of the meaning and use of messages.

Term	Definition/Explanation
Communication Gap	Refers to the difference between what was communicated by the extension agent & what has actually been received by the audience.
Community	Refers to groups of mutually dependent people, living in a more or less compact continuous geographical area, having a sense of belonging & sharing common values, norms & some common interests & acting collectively in an organized manner to satisfy their chief needs through a common set of organizations & institutions (Chitambar).
	Community is sub-group having many of the characteristics of society but on a smaller scale and with less intensive and coordinated common interest.
Continuous Education	Continuous education can be given through both formal and informal types of education as and when need arises to educate for new technological development and for socio-cultural development.
Contract Farmer	A farmer who act as intermediary between follower farmers and the extension education agency.
Co-operatives	These are the institutions organized by groups of individuals for mutual help, thrift and, benefit
Culture	Is that complex whole which includes knowledge, belief, and art? Morals, law, custom & other capabilities & habits acquired by the people as member of society.
	Culture is the sum total of the ways in which human beings live and is transmitted from generation.
Customs	Accepted ways of meeting marriage, eating, death and birth.
Decision Making	Is the process of consciously choosing courses of action from available alternatives & integrating them for the purpose of achieving the desired goal?
	Decision making refers to the process in which programme are prepared by various departments in consultation with various person. .

Term	Definition/Explanation
Demonstration	Demonstration is conducted by farmer under direct supervision of extension workers to prove the advantages of recommended practices.
Demonstration Centers	These provide effective communication about improved methods of farm practices to farmers by practical displays.
Demonstration Plot	It is a small area set aside for result demonstration of extension staff.
Diffusion	Is the process by which an innovation is communicated through certain channels over time among the members of a social system.
Distance Education	Is the process of educating large number of people, dispersed & distantly located, with little face-to-face interaction between the teacher & the taught?
	It is anyone of the various forms of the study which are not under the continuous immediate supervision of tutors present to their students in lecture rooms.
Dramatics	Dramatics are a great source of entertainment and education and have been known to have a considerable effect on the society form times immemorial.
Education	It is the process of developing capabilities of the individuals so that they can adequately respond to their situations.
Emotions	Denote a state of being moved, stirred up or aroused & involve impulses, feeling & physical & Psychological reactions.
Empathy	Is the ability of an individual to project oneself into the role of another person, to be able to appreciate the feelings, thinking & actions of another person?
Esteem	The evaluation of an individual's role or behaviour in status which he occupies is called esteem.
Ethnocentrism	Is the tendency of the people to consider their own culture of high value & superior to all others & is found in cultures all over the world.
Evaluation	Evaluation is the study of the degree of success, we get in terms of our original objectives of the programme.

Term	Definition/Explanation
Evaluation in Extension	Is used to determine whether an extension programme has achieved its goals & whether these goals could have been achieved more effectively in a different way. It enables extension administrators' managers & agents to learn more effectively from their experience by systematic observation & analysis of this experience. (Van den Ban & Hawkins).
Exhibition	An exhibition is a systematic display of models, specimens, charts, photographs, pictures, posters, information etc. in a sequence around a theme to create awareness & interest in the community.
Extension	Is an education in agricultural & in home economics for rural People, aimed at improving farm & home. (Blackburn & Flaherty).
Extension Agent	The extension worker at the village level in direct day to day contact with local farmers.
Extension Education	Is an applied Science consisting of content derived from research, accumulated field experiences & relevant principles drawn from the behavioural Sciences synthesized with useful technology into a body of philosophy, principles, content & methods focused on the problems of out–of–School education for adults & youth. (Leganes)
Extension Worker	Men and women employed by an extension agency or technical and professional staff employed at all level.
Family	Is a group defined by a sex relationship sufficiently precise & enduring to provide for the procreation & upbringing of children (MacIver & Page).
	Family is a system of organized relationship involving workable and dependable ways of meeting basic social needs. It is a most multifunctional of all institutions in society.
Farm Clinic	Is a facility developed & extended to the farmers for diagnosis & treatment of farm problems & to provide some specialist advice to individual farmers.

Term	*Definition/Explanation*
Farming Need	Farming needs are the gap between farming practices adopted by the farmers at present and farming practices recommended by the research worker at present.
Feedback	Means carrying some significant responses of the audience back to the communicator. Extension communication is never complete without feedback information.
Fidelity	Is the faithful performance of communication process by all its elements viz. Communicator, message, channel & receiver.
Field Day	Is a method of motivating the people to adopt a new practice by showing what has actually been achieved by applying the practice under field conditions? A field day may be held in a research farm or in a farmer's field or home.
Film Strips	These are the series of still picture in one roll.
Films	Films are a series of motion pictures projected in sequences with a synchronized sound system.
Flannel Graph	It. is piece of flannel or Sand paper stick on funnel.
Flash Cards	These are small cards of thick paper with pictures, diagrams or drawings on a particular subject or topic.
Folder	It is a single printed sheet of paper of big size folded once or twice & gives essential information relating to a particular topic.
Folk Songs	Folk songs are a powerful means of propagating the idea and experience of a wise and learned.
Folk Ways	Folk ways are customary ways of behaving in which society exerts somewhat, they are not rigid.
Formal Education	Are the highly institutionalized, chronologically graded & hierarchically structured education system, spanning lower primary school & the upper reaches of the university?
Forum	Two or more speakers who present talks on the same subject. The primary characteristics of the forum are that the subject is controversial and speakers present the opposite side. of question.

Term	Definition/Explanation
Frustration	Is a condition in which an individual perceives the wish goal blocked or unattainable, which creates some tension in the individual?
Functional Literacy	Functional literacy is the ability to read, write and calculate, so that person may engage in all.
Goal	Goal is the distance in any given direction proposed to be covered in a given period of time.
Graph	Graph is a diagrammatic presentation of quantitative data in a clear, interesting and easily understandable form and is very effective in showing trends, developments and relationships.
Group	A group is a unit of two or more people in reciprocal communication & interaction with each other.
Group Meeting	Is a method of democratically arriving at certain decisions by a group of people, by taking into consideration the members' points of view?
	Group meeting are a device to save time by several individuals simultaneously.
Homophily – Heterophily	Is the degree to which pairs of individuals who interact are similar in certain attributes, such as beliefs, values, education, social status & the like? On the other hand, Heterophily is opposite of homophily & is defined as the degree to which pairs of individuals who interact are different in certain attributes.
Illiteracy	Illiteracy means inability to read and write.
	It means person who has not acquired basic skills in reading, writing or arithmetic.
Individual Meeting	These are between two or three individuals arranged at home or at the field.
Informal Education	Is the lifelong process by which every person acquires & accumulates knowledge, skills, attitudes & insights from daily experiences & exposure to the environment at home, at work, at play etc.
Innovation	Is an idea, practice or object that is perceived as new by an individual or other unit of adoption.
	Innovation is an idea perceived as new by the individual.

Term	Definition/Explanation
Innovation-Decision Process	Is the process through which an individual or other decision-making unit passes from first knowledge of an innovation, to forming and attitude toward the innovation, to a decision to adopt or reject to implementation of the new idea, & to confirmation of this decision?
Innovativeness	It is the degree to which an individual is relatively earlier in adopting new ideas than the other members of his, social system.
Institution	Institution is a system of relationships or a pattern for carrying out an idea or a desire which is regarded as necessary for the welfare of the group.
Integrated Rural Development	IRD is defined as development which can help to increase the purchasing power of the rural poor through the generation of greater opportunities of gainful employment,
Interest	Interest refers to the attention with a sense of concern focused on some objects.
Journal/Magazine	These are periodicals, containing, information related to various topics of interest not only for the farmers but also for the extension agents.
Leader	An individual is leader in any social situation in which his ideas and action influence the thoughts and behavior of others.
Leadership	Within the power structure of every society certain vital, integral, individuals operate within groups to promote, stimulate, guide or otherwise influence members to action such activity has been called leadership, & the individuals has been referred to as leaders. Leadership is a process of influencing the behaviour of the individual in a real situation.
Leaflet	It is a single printed sheet of paper of small size, containing preliminary information relating to a topic. Leaflet is a single sheet of printed matter folded or unfolded giving information on a particular topic.
Learning	Is the process by which an individual, through one's own efforts & abilities changes the behavior?

Term	Definition/Explanation
Learning Experience	It is a mental Physical reaction, one makes through seeing, hearing or doing the things to be learnt through which one gains meaning and understandings in solving new problems.
Mass Meeting	These are the meetings of a general mass people called to hear a leader or endorse the decisions of an executive committee, formed by the general body of an organization.
Message	Is the information related to the recommendations from research, the technology, constitute the content or subject matter, which is relevant to a particular set of audiences. A good message should clearly state what to do, how to do, when to do & what would be the result.
	It is the information, a communicator wishes his audience to receive, understand, accept and act upon.
Method Demonstration	Is given before a group of people to show how to carry out an entirely new practice or an old practice in a better way. It is essentially skill training, where the emphasis is on effectively carrying out a job, which shall improve upon the result. Method demonstration is sometimes used as compel.
	Method demonstration are those in which a new method, or technique is demonstrated, taught or given practice of.
Models	It is the replicate of real objects or sealed representation of things.
Monograph	Monographs are a technical publication, written on a particular subject.
Mores	Are socially acceptable ways of behavior that do involve moral standards? For example, rearing of a particular animal, in spite of its profitability, may not be acceptable to a group or society.
Motivation	It is a need satisfying & goal seeking behavior which includes drives, desires, needs & similar forces (Wilson & Gallup).
	Motivation is the process of creating a strong desire for a appealing to the fundamental wants of the people so that they are influenced to change the behaviour.

Term	Definition/Explanation
Need	It is a gap between what is, the existing situation, & what ought to be, the desirable situation.
	Norms refers to the rules that govern action directed towards achieving values.
Newsletter	It is a miniature newspaper in good quality paper, containing information relating to the activities & achievements of the organization. It has a fixed periodicity of publication.
Newspaper	Is a bunch of loose printed papers properly folded, which contains news, views, advertisements etc. & is offered for sale at regular intervals, particularly daily or weekly?
Newspapers	Newspapers are an effective medium for disseminating information about local, district, provincial, country and world affairs.
Non–Formal Education	Is an organized, systematic, educational activity carried on. Outside the frame work of the formal system to provide selected types of learning to particular sub groups in the population, adults as well as children according to their needs, e.g., Agricultural extension?
Norms	Are the established behaviour Patterns for the members of a social system. An individual who first adopts a new technology may be regarded as a deviant by others.
Objective	Objective is the direction of movement.
ON-Farm Testing	These are conducted on the farmers' field on such problems where the appropriate technologies are not available for particular agro-climatic situation to transfer, & the relevant research information available does not suit the situation from the point of view of the farmers.
Opinion Leaders	Those individuals from whom others seek the information and advice.
Organization	Organization is a set of two or more parts which are so related that they function, as a whole.
Over adoption	Sometimes it may happen that people continue to adopt an innovation, rather vigorously, when experts feel that it should not be so done called as an over adoption, e.g., Excessive use of pesticides.

Term	Definition/Explanation
Overtime	Over time refers to the time required to adopt from its origin.
Pamphlets	Pamphlets are very valuable and effective written material or literature for, use in extension.
Panchayati Raj	Panchayati Raj refers to members or an assembly of elders who settles disputes. According to custom in a village or within a caste.
Panel	Panel is also termed as round table discussion. It usually composed of 4-6 persons. The members of the panel carry on a discussion of the topic among themselves and the 'audience listens in on the discussion.
Perception	Is the process by which an individual receives stimuli through the various senses & interprets them? Is an activity through which an individual becomes aware of objects around & of events taking place?
Personal Example	It is one of the most effective ways of training and communicating with the people. It is generally what we do and how we do it makes a lasting impression on trainer or people.
Personality	Is the unique, integrated & organized system of all behaviour of a person, includes physical traits, attitudes, habits, emotional & psychological characteristics. Personality is a pattern body of inherited traits.
Persuasion	It is more democratic in influencing the audience to bring about change in their attitude & behaviour. In persuading people, the extension agent provides lots of arguments in flavour of acceptance of the recommendations & provides evidences of gain.
Philosophy of Extension	Is to teach people how to think not what to think (Kelsey & Hearne)
Plan of Work	It is an outline of activities so arranged as to enable effective execution of the entire programme.
Planning	Planning is a process of preparing systematic statement of the line of action for achieving certain overall and specific objectives in relation to needs and resources.

Term	Definition/Explanation
Poster	Poster is an important visual aid helping to arouse people.
Poverty	Is a condition in which a person is not capable of procuring the minimum amount of food required for ones own sustenance & of the members of the family In India, it is about 2250 Kcal per day for an individual.
	Poverty is the social phenomena in which a section of the society is unstable to fulfill even its basic necessities of life.
Prejudice	Is judgment before due examination & consideration of facts, based on certain assumptions generally lead to the formation of prejudice? Prejudice is normally negative & difficult to reverse.
Prestige	Prestige is the evaluation of the status.
Programme	Programme is a systematic statement of situation, objectives, problems and solutions. It is relatively permanent but requires constant revision.
Programme Planning	Is a decision making process involving critical analysis of the existing situation & the Problems, evaluation of the various alternatives to solve these problems & the selection of the relevant ones, giving, necessary priorities based upon local needs & resources by the cooperative efforts of the people both & non-official with a view to facilitated the individual & community.
Project	Project refers to an outline of procedure and portions to some phrase of extension work.
Propaganda	Is deliberate manipulation of people's beliefs, values & behaviour through words, gestures, images, thoughts, music etc. propaganda aims at propagating beliefs & values of the propagandist & presents only communicators side of arguments without considering the arguments of the receivers side.
Psychology	Is the science of human behaviour.
	It refers to the use of animals in behavioral research and collection information and in sight relevant to their area of concern.

Term	Definition/Explanation
Publicity	Is based on truth & propaganda often suppresses the truth. One sided communication giving view-points of only the message-source, ignoring the view-points of receivers of the message, may sound propaganda despite the message being based on truth.
Reference Group	Reference group of persons whom an individual consult before taking an important decision.
Religion	Human resources to the self of belief and feeling of the presence of god to understand human mankind.
Result Demonstration	Is a method of motivating the people for adoption of a new practice by showing its distinctly superior result? The demonstrations are conducted in the farm or home of selected individuals & are utilized to educate & motivate groups of people. This is a very effective method for the transfer of technology in a community. These are those in which two practices or techniques are compared for results.
Role	Role refers to the function of the status.
Rural Sociology	Rural Sociology is the systematic study of the people living in rural areas and follow patterns of occupation and life somewhat different than those living in urban areas. Sociology is the science of human relationships. Rural sociology involves the study of human relationships in rural situations.
Rural Development	Rural development is a process of change from the traditional way of living of progressive way of living.
Social Centre	The adult education classes will be organized through different adult education centres on the basis of needs and requirements of adult men and women.
Social Class	Social class is an abstract category of persons arranged in levels according to the social status they posses.

Term	Definition/Explanation
Social Control	Is the pattern of influence the society exerts on individuals & groups to maintain order & establish rules in the society.
	Social control is the process of constraining persons to conform to various norms.
Social Group	Social group is a unit of two or more persons in reciprocal interaction or communication with each other.
Social Institutions	Is an organized system of social of relationships which embodies certain common values & procedures & meets certain basic needs of the society? (Horton & Hunt).
Social Interaction	Is dynamic interplay of forces in which contact between persons & groups results in a modification of the attitudes & behaviour of the participants. (Sutherland).
Social Problems	It is a condition society that is considered harmful or undesirable by society as a whole based on existing social values.
Social Stratification	Is the arrangement of individuals or groups of people into hierarchically arranged strata in a community?
	Social sanction is a regard or punishment applied to person in order to encourage or discourage certain types of behaviour.
Social System	Is a set of interrelated units that are engaged in joint problem solving to accomplish a common goal (Rogers).
	A population of individuals who are functionally differentiated and engaged in collective problems and solving the problems.
Social Value	These are defined as the attitudes held by individuals, group or society as a whole as to whether material or non material objects are good, bad, desirable or undesirable.
Social Welfare	It refers to large number of activities connected with the uplift and welfare of the weak and exploited sections of the society.

Term	Definition/Explanation
Society	Is a group of people who have lived together long enough, sharing common values & general interests, to be considered as a social unit? Society refers to the group within which men share.
Sociology	Sociology is the scientific study of human's behavior in relation to other. Groups and individuals with whom he interacts.
Sociometric Technique	It is a Scientific method for identification of leaders & leadership pattern.
Specimen	It may be part of an object or one of group or class to represent the whole group.
Staff Meeting	These meetings are among formal groups' staff meeting occupying a unique position. They are an essential part of successful extension programme.
Status	It is a position or place in a social set-up.
Stereotype	Are fixed images farmed in one's mind about people, Practices or various other social phenomena on the basis of experience, attitudes, values, impressions or without any direct experience?
Study Tour	A group of interested persons accompanied & guided by one or more extension agents moves out of their neighborhood to study & learn significant improvements in farm & home elsewhere. The main purpose is to motivate the visitors by showing what others have been able to achieve.
Supervision	It is the process of placing into action standards of performance developing and utilizing human resources to achieve organization results.
Sustainability	Is the successful management of resources to satisfy changing human needs, while maintaining are enhancing the quality of environment & conserving natural resources?
Teaching	Is the process of arranging situations in which the important things to be learned are called to the attention of the learners, their interest developed, desire aroused & action promoted (Leganes). Teaching is the process of arranging situations that stimulate and guide learning activities in order to bring desired changes in human behaviour.

Term	Definition/Explanation
Technical Workers	These workers are trained in special fields. Their training includes some basic training in. specialized fields followed by job training.
Television	Television is one of the most .powerful audio visual aids. It combines in itself both visual and auditory aids and is very effective in conveying actual life events in action.
Time Charts	They present data in ordinary sequence, e.g., year to year increase in the income of the people, profit and loss account of farms etc.
Traditions	The uniformity accepted ways of thinking.
Treatment of message	Treatment means the way a message is handled, dealt with, so that the information gets across to the audience.
Values	As conception of the desirable, as standards of evaluation, as guides for decision- making behaviour or simply as expression of Preference (Kahl).
Village Level Worker	The multipurpose agent of India's Community development programme at village level.
Visual Aid	These are aids which facilitate understanding of a problem or topic through seeing. The instructional devices through which the message can only be seen are known as visual aids, e.g., Photographs.
Walking Tours	Walking tours are an important means of communicating with people.
Wall newspaper	These are big sheets of paper with current news, experiences, and recommendations etc., printed on them.
Wish	It is a common pattern of human behaviour involves hopes for future achievement. Such ambitions & goals are generally termed as wish.

IMPORTANT FACTS/POINTS TO BE REMEMBER

1. A decision is a choice among alternatives.
2. A good extension programme should be flexible.

3. A good extension service is a process of helping the people to help themselves.

4. A new concept of budgeting is performance budgeting, the introduction of which has been recommended by the Administrative Reforms Commission.

5. A prepared script should be divided into different parts depending upon the nature of the message to be conveyed.

6. A screen is to provide the maximum possible visibility of the image with minimum glare or strain on the eyes. A single communication model consists of Source(S), Message (M), Channel (C), Receiver (R) and Effects of communication (E).

7. A tape stored under normal room temperature and humidity is not erased, but holds sharp and clear quality for many years.

8. Actions follow when desire, conviction and the prospect of satisfaction make it easier for the person to act rather than not to act.

9. Administration refers to the guidance, leadership and control of efforts of a group of individuals towards some common goal.

10. Aim of extension work is to bring about the changes towards betterment.

11. All the rural listeners' may or may not be educated in formal sense.

12. An exhibition is a systematic display of models, specimens, charts, posters, information etc., in a sequence around a theme to create interest in the community.

13. An extension worker is also learning, while giving the learning experience to the people.

14. An extension worker should utilize the latent goodwill of the people in extension programmes.

15. An innovation is idea, practice or object perceived as new by an individual.

16. Associations are organized for particular purposes.

17. Attitude denotes the degree of positive or negative feelings associated with some psychological object towards which people differ in varying degrees.

18. Authority in an organization is the power in a position to exercise discretion in making decisions affecting others.

19. Basic education is designed to provide literacy, an elementary understanding of science and environment.

20. Campaign denotes an intense educational activity for motivating and mobilizing a community to action, to solve a problem or satisfy a need urgently felt by it.

21. Committee is a small group of person to who some matter is referred to mainly for detailed examination and decision.

22. Contingency planning denotes planning for possible future environments which are not expected to occur, but which may occur; if this possible future is widely different from that premised, alternative premises and plans are required.

23. Culture is defined as the socially standardized way of feeling.

24. DASP stands for Diversified Agriculture Support Project.

25. 'DIORAMA' is scenic representation of the original, with specimen, model, and painting.

26. Drive is defined as a tendency initiated by shifts in the physiological balance in the body.

27. Economic development is accorded priority in programmes of rural development.

28. Education is the production of desirable changes is human behaviours.

29. Education should be conceived as a lifelong process of learning.

30. Efficiency-cum-performance audit is a technique of audit adopted to assess and evaluate the economy, efficiency and effectiveness of development projects in an organization.

31. Empathy is the capacity to see one self in other fellow's situation.

32. Evolution should be in line with the overall purposes of the agency.

33. Exhibition, Radio and Literature are institutional communication methods.

34. Extension education is the education for rural people outside the regularly organized schools and class rooms to bring about social and cultural development.

35. Extension has to carry research findings to the rural people and establish their relevance in fulfilling their needs.

36. Extension helps in adoption of innovations.

37. Extension programme is a statement of situation, objectives, problems and solutions.

38. Extension programmes should be people programmes with government aid.

39. Extension work cannot thrive without trained specialists.

40. Extension workers with superior knowledge and wisdom can help in solving problems of rural society and lead them on the voyage to modernization.

41. Facts are the foundation stone on which the community leaders and the problem committee build and carry out their programmes.

42. Farmers training centres imparts training to farmers, farmers leader, rural youths and farm women.

43. Films are one of the most effective means of arousing interest.

44. Formal education is highly institutionalized, chronologically graded and hierarchically structured to being with primary school upto university education.

45. Friends and neighbors are first in importance in the evaluation, trials and adoption stages of innovation(s).

46. Heterophily is the degree to which pairs of individuals who interact are different in certain attributes.

47. Home is the fundamental unit of civilization; family is the first training group of human race.

48. Homophily is the degree to which pairs of individuals who interact are similar in certain attributes, such as beliefs, values, education, social status and the like.

49. I.V.L.P. is basically a linkage programme.

50. If a child of 8 can do all tests meant for the 10 years old then his I.Q. will be 125.

51. Imagination will put new glitter on old things and will lead to seek new and better view points.

52. In extension programme people's participation is based on self help and self confidence.

53. In extension, educational psychology change agents are more concerned with the adult or grown up people.

54. In presentation of a talk just talk to the people and do not speak naturally.

55. In some cases use of progressive leader farmers is made to produce the multiplier effect.

56. In the earlier days rural development was considered synonymous with community development.

57. In the sound track, sound is photographed on the photographic film.

58. Innovativeness is the degree to which an individual is relatively earlier in adoption of new ideas than others of his social system.

59. Intelligence quotient is equal to mental age divided by chronological age and multiplied by 100.

60. It is necessary to base the extension education programmes on the psychology of the rural people to make them effective.

61. Judgment is an assignment of value to alternatives.

62. Khetihar Mazdoor Bima Yojna was started on 2001-02.

63. 'Kisan Bharti' periodical is published from Pant Nagar.

64. Leading is the process of influencing people so that they will strive willingly and enthusiastically toward attainment of the organizational goals.

65. Learning calls for effective communication.

66. Learning is affected by physical and social environment.

67. Literature is the basic of any teaching programme.

68. Method of mass communication is Newspaper, Television and Radio.

69. Motivation is energy mobilization towards the attainment of goals.

70. N.A.T.P. stands for National Agriculture Technology Project.

71. NARD stands for National Fund for Rural Development.

72. National Academy of Agriculture Research Management is located at Hyderabad.

73. Need and interest based programmes yield better results than the imposed ones.

74. Neglect of one member of a family by extension worker may lead to rejection of an innovation.

75. Next to camera and film the most important ingredient for picture-taking is a good imagination.

76. Non-formal education denotes any organized, systematic educational activity carried outside the framework of the formal system.

77. Objective of Antyodaya Yojna (1977-78) is to make the poorest families of the village economically independent.

78. Objectives present the directions of movement, while a goal is the distance in a given direction one expects to go during a given period of time.

79. Personality is the dynamic organization within the individual of those pshycological systems that determine the unique adjustments to his environments.

80. Philosophy is a body of general principles/laws as a field of knowledge.

81. Plan of work comprises a definite-outline of procedure for solving the different problems of an extension programme.

82. POSDCORB denotes planning, organization, staffing, directing, co-ordination, reporting and budgeting.

83. Psychology motive refers to some internal activator within an individual.

84. Radio belongs to the spoken word means of communication.

85. Reference group is a group of persons whom an individual consults before taking an important decision.

86. Results of programme evaluation/assessment greatly help in improving the quality of programme(s).

87. Rural development is the result of many interacting forces and education is one of them.

88. Rural development programmes should be form to meet short term change.

89. Rural society is more homogeneous in nature, relatively independent and with a low degree of social differentiation.

90. Rural sociology is the systematic study of people living in rural areas and who are living by or are immediately dependent on agriculture.

91. Sampurna Gramin Rojgar Yojna (2001) provides employment and food security.

92. Socialization is the process of inducting the individual into the social and cultural world to play his role as a member of society.

93. Spontaneity and creativity are important parts of brain storming.

94. Successful teaching contributes to effective learning.

95. The ability and capability of an individual should be brought out and sharpened by an extension worker.

96. The advantage of the over head projector is that a teacher can always face the class.

97. The basic personal skills needed for effective administration are technical, human and conceptual.

98. The basic philosophy of extension is directed towards changing the outlook of persons by educating them.

99. The education designed to develop a specific knowledge and the skills associated with economic activities beneficial to make respectable living is called occupational education.

100. The extension work must be based on the need and interest of people.

101. The extension worker should understand the basic wants and incentives of the people with he/she has to work.

102. The extension workers should have faith in democratic values and should try to educate people.

103. The family is the primary institution of society.

104. The first principle of extension service rests on, "citizen is sovereign in the democracy".

105. The fundamental objective of extension education is development of people.

106. The leadership qualities are developed in rural people, if they participate in extension programmes.

107. The main activities of Sevagram Attempt comprised organization of training centres for cottage industries, prohibition and removal of untouchability.

108. The means by which sound is reproduced from a sound motion picture film is practically the reverse of the way it is recorded.

109. The members of a secondary group have little personal affection and their relationship is governed by the fulfillment of some objective.

110. The need of training is to develop knowledge, skill and attitude.

111. The participation of the people in the programme planning process normally gives them a proprietary interest in seeing that an action programme.

112. The persons in the primary group have face-to-face relationship, such as a family and play group.

113. The principle of 'One village, One society' was suggested by Mclang Committee.

114. The rural people should voluntarily participate in the extension work.

115. The Sevagram Attempt was started under the guidance of Mahatma Gandhi in 1920.

116. The success of the extension work lies in satisfaction participation of the people.

117. The technical messages are carefully selected and are relayed during the visit to the farmers.

118. The training and visit system comprises an excellent example of the training approach.

119. The training approach emphasizes more systematic and deeper learning of specific basic skills and related knowledge.

120. Working objectives are the specific items which are to be achieved.

2

ABBREVIATIONS

Abbreviation	Full Form
@	At
m	micro
mm	micromole
A	Ampere
A.M.	Time from midnight 12 to Noon 12
a/c	Account
A+bgr	Above and below ground
A°	Angstrom
AARDO	Africa-Asian Rural Development Organization
AARRO	Afro-Asian Rural Reconstruction Organization
AAU	Assam Agricultural University
ABC	Accuracy, brevity & Clarity of News Story
Abe	Abequose
AC	Agricultural College
ACIAR	Australian Centre for International Agricultural Research
ACRBCS	Advanced Center for Research on Black Cotton Soils of Karnataka
ACT	Advanced Centers of Training
ACT	Audio Cassette Technology
ADB	Asian Development Bank
ADI	Acceptable Daily Intake
AEO	Agril Extension Officer
AEZ	Agricultural Export Zones
AFLP	Amplified Fragment Length Polymorphism
AFRC	Agricultural Food Research Council
AGRIS	International Information System for Agricultural Sciences and Technology

Abbreviation	Full Form
AHRD	Agricultural Human Resource and Development
AHRD	Agricultural Human Resource Development Project
AIC	Agriculture Insurance Company
AICC	All India Congress Committee
AIP	Agribusiness and Innovation Plat Form
AIR	All India Radio
AKIS	Agricultural Knowledge & Information System
AMU	Aligarh Muslim University
AOA	Agreement on agriculture
APAU	Andhra Pradesh Agricultural University
APC	Agricultural Price commission
APC	Armored Personnel Carrier
APEC	Asia Pacific Economic Cooperation
APEDA	Agricultural & Processed Foods Product Export Development Authority
APEDA	Agricultural and Processed Food Products Exports Development Authority
Ara	Arabinose
ARDC	Agricultural Refinance & Development Corporation
ARIC	Agricultural Research Information Centre
ARIS	Agricultural Research Information System
ARS	Agricultural Research Services
ARS	Agriculture Research Service
ASCIT	American Standard Code for Information
ASRB	Agriculture Service Requirement Board
ATIC	Agricultural Technology Information Centre
Atm	Atmosphere
ATM	Automatic Trained Machine
ATMA	Agricultural Technology Management Agency
ATP	Adenosine Triphosphate, the currency of energy.
AUD	Ambedkar University, Delhi

Abbreviation	Full Form
AVA	Audio-Visual aids
AVRDC	Asian Vegetable Research and Development Center
BAC	Bihar Agricultural College, Bhagalpur
BAR	Bhabha Atomic Research Centre,Trombay, Bombay
BAU	Birsa Agricultural University
BCKVV	Bidhan Chandra Krishi Vishwa Vidyalaya
BCSBI	Banking Code of Standard Board of India
BHU	Banaras Hindu University
BLAST	Basic Local Alignment Search Tool
BMB	Bhartiya Mahila Bank
BMB	Biochemistry and Molecular Biology
BNHS	Bombay National History Society
Bq	Becaquerel
BRICS	Brazil, Russia, India, China and South Africa
BSA	Bovine Serum Albumin
BSI	Botanical Survey of India
BTU	British Thermal Unit
BVC	Bihar Veterinary College
c	centi
C	Coulomb
C°	Degree Celsius
CA	College of Agriculture
CAB (I)	Commonwealth Agricultural Bureaux (International since 1986)
CAD	Command Area Development Programme (1974-79)
CAD	Computer aided design
CADA	Command Area Development Authority
CAG	Central Agency of Government or Comptroller and Auditor General (CAG) of India
Cal	Calorie
CAM	Crassulacean Acid Metabolism
CARI	Central Avian Research Institute, Izatnagar.

Abbreviation	Full Form
CARI(AN)	Central Agricultural Research Institute for Andaman and Nicobar Group of Islands
CARIANGI	Central Agricultural Research Institute for Andaman and Nicobar Groups of Islands, Port Blair, Andamans and Nicobar 744 101
CAZRI	Central Arid Zone Research Institute
CBI	Central Bureau of Investigations
CBRI	Central Bee Research Institute
CCEA	Cabinet committee on economic affairs
CCIRG	Climate change impart review group
CCP	Critical control point
CCPA	Cabinet committee on political affairs
CD	Compact Disc
C-Deck	Center for Advanced Computer
CDP	Community Development Programme (1952)
CEA	Controlled Environment Agriculture
CENATEL	Center national de teledetection et de surveillance due couvert forestier/National center for remote sensing and forest cover monitoring
CeRA	Consortium for e-Resources in Agriculture or Cambridge Energy Research Associates or Continuous erythropoietin receptor activator.
CERS	Certified emission reduction
CET	Central Eligibility Test
CF	College of Fisheries.
CFC	Chloro Fluoro Carbon
CFTC	Commonwealth Fund for Technical Co-operation
CFTRI	Central Food Technological Research Institute
CGH	Comparative genomic hybridization.
CGIAR	Consultative Group of International Agricultural Research
CGPCS	Contact group on piracy off the coast Somalia
CGRI	Central Goat Research Institute
CGTMSE	Credit guarantee fund trust for micro and small enterprises

Abbreviation	Full Form
CHES	Central Horticulture Experiment Station
CHS	College of Home Science
Ci	Curie
CIAE	Central Institute of Agricultural Engineering
CIAT	Centro International de Agriculture Tropical (International Center for Tropical Agriculture) Coli-Columbia
CIBA	Central Institute of Brackishwater Aquaculture
CICFRI	Central Inland Capture Fisheries Research Institute, Barrackpore
CICR	Central Institute of Cotton Research
CIDA	Canadian International Development Agency
CIFA	Central Institute of Freshwater Aquaculture
CIFE	Central Institute of Fisheries Education
CIFRI	Central Inland Fisheries Research Institute
CIFT	Central Institute of Fisheries Technology
CIHNP	Central Institute of Horticulture for Northern Plains
CIMAP	Central Institute of Medicinal and Aromatic plants
CIMMYT	Centro International de Mejorimiento de Maiz y Trigo (International Centre for the Improvement of Maize and Wheat)
CIMMYT	International Center for Maize and Wheat Improvement
CIP	Centro International de la Papa (International Potato Centre)
CIPET	Central Institute of Post-harvest Engineering and Technology
CIRB	Central Institute for Research on Buffaloes
CIRG	Central Institute for Research on Goats
CIS	Crop Insurance Scheme
Cm	Centimeter
CMFRI	Central Marine Fisheries Research Institute
CMRS	Central Mango Research Station
CNG	Carbon free Natural Gas
CNG	Compressed natural gas
CO	Communication Office
COD	Chemical Oxygen Demand

Abbreviation	Full Form
COP	Conference of parties
COR	Corundum
CP	Cast polypropylene
CPCRI	Central Plantation Crops Research Institute
CPD	Co-ordination and Publication Division
Cpm	Chemical vapour deposition
CPPTI	Central Plant Protection Training Institute
CPRI	Central Potato Research Institute
CPSE	Central public sector enterprise
CRIDA	Central Research Institute for Dryland Agriculture
CRIJAF	Central Research Institute for Jute and Allied Fibers
CRRI	Central Rice Research Institute
CSAUAT	Chandra Shekhar Azad University of Agriculture and Technology
CSIR	Council of Scientific and Industrial Research
CSR	Council of Sugarcane Research
CSSRI	Central Soil Salinity Research Institute
CSWCRTI	Central Soil and Water Conservation Research and Training Institute
CSWRI	Central Sheep and Wool Research Institute
CTCRI	Central Tuber Crops Research Institute
CTRI	Central Tobacco Research Institute
CTRL	Cotton Technology Research Laboratory
CVS	College of Veterinary Science
CVSAH	College of Veterinary Science and Animal Husbandry
d	deci
D.Sc	Doctor of Science
Da	Dalton
DARE	Department of Agricultural Research & Education
DBT	Direct benefit transfer scheme
DC	Dryland Cereals

Abbreviation	Full Form
DCA	College of Agriculture
DDA	Djidja-Dan-Atcherigbe
DDG	Deputy-Director-General
DDT	Dichlorodiphenyl trichloroethane
DEC	District Extension Centre
DG	Director General
DIGE	Differential gel electrophoresis
DLARP	Dryland Agricultural Research Projects
dm	Decimeter
dm	Dry matter
DMR	Directorate of Millet Research
DOM	Dead organic matter
DOR	Directorate of Oil
DOS	Disk operating system
DPAP	Drought prone Area Programme
DPD	Date Processing Division
dpm	disintegrations per minute
dps	disintegrations per second
DRR	Directorate of Rice Research
DSR	Directorate of Sorghum Research
DTC	Direct Tax Code
DTH	Direct To Home
DWCRA	Development of women and children in Rural Areas
EC	Electrical Conductivity
EC	Empowered Committee
ECA	Economic Commission for Africa
ECE	Economic Commission for Europe
ECWA	Economic Commission for Western Asia
EEC	European Economic Community
EEI	Extension Education Institutes

Abbreviation	Full Form
EFDB	Emission factors data base
EFL	Ecologically fragile lands
EFP	Joint UN/FAO World Food Programme
EGC	Engine Gear and Clutch
EIA	Environment Impact Assessment
EIT	Economy in transition
ELISA	Enzyme Linked Immune Sorbent Assay
EM	Emerging Market
EMI	Equated Monthly Installment
EMS	Environment Management System
Encl	Enclosure
ENSA	Ecole National Superieur Agronomique
EPA	Environment Protection Agency
EPP	Express Post Parcel
ESCAP	Economic and Social Commission for Asia and Pacific
ESI	Electro-Spray Ionization
ETC	Extension Training Centers
EVMs	Electronic voting machines
F	Faraday
f	Femto
F.O.R.	Free of Rail charge
FAO	Food and Agricultural Organization
FCCC	Framework Convention on Climate Change
FD&C	Food, Drug and Cosmetics
FDA	Food and Drug Administration
FDI	Foreign Direct Investment
FETS	Farm, Engineering & Transport Services
FIG	Farmers Interest Group
FIR	First Information Report/First Investigation Report
FISH	Fluorescent-*in-situ* hybridization

Abbreviation	Full Form
FLD	Front Line Demonstration
FM	Frequency modulation
FMU	Field Medical Unit
FOD	Field Operations Division (FOD)
FRA	Forest Resources Assessment
FRI	Forest Research Institute
FRS	Fruit Research Station
Fru	Fructose
FS	Financial Services
FSRE	Farming Systems Research & Extension
Ft	Feet
FWP	Food For Work Programme
G	gauss
G	giga
g	gram
GA	General Assembly
GABA	g-Aminobutric acid
Gal	Galactose
GalN	Galactosamine
GATT	General Agreement on Tariffs and Trade
GAU	Gujarat Agricultural University
GBPUAT	Govind Ballabh Pant University of Agriculture and Technology
GBq	gigabecquerel
GC-MS	Gas chromatography-Mass spectrometry
GDP	Gross Domestic Product
GFP	Green Fluroescent Protein
GHE	Green house effect
GHG	Greenhouse gas
GIS	Geographical Information System
GJEPC	Gems and jewellery export promotion council

Abbreviation	Full Form
GL	Grain Legumes
GlaNAc	N-Acetylgalactosamine
GleA	Glucuronic acid
GleN	Glucosamine
GleNAc	N-Acetylglucosamine
Glu	Glucose
GMO	Genetically Modified Organisms
GNI	Gross national income
Gold	Genomes online databases
GOM	Group of ministers
Govt.	Government
GP	Gypsum Requirement
GPC	Good practice guidance
GRAS	Generally regarded as safe
GSLV	Geosynchronous Satellite Launch Vehicle
GST	Glutathione-S-transferase
GTC	Gramsewak Training Centers
GVM	Global vegetation model
GWP	Global warming potential
h	hour
ha	hectare
HAU	Haryana Agricultural University
HFS	Housing and Food Services
HI	Harvest Index
HID	High intensity discharge
HPKVV	Himachal Pradesh Krishi Vishwa Vidyalaya
HRD	Human Resource Development
HRS	Human Resources Services
HSRP	High Security Registration Plate
Html	Hyper Text Markup Language

Abbreviation	Full Form
http	Hyper Text Transfer Protocol
HYTECH	High tech Seed India Pvt. Ltd.
HYV	High Yielding Varieties
HYVP	High yielding variety programme
IA	Internal Audit
IAAP	Intensive Agricultural Area Programme
IACD	ICRISAT Association for Community Development
IADP	Intensive Agricultural District Programme
IAEA	International Atomic Energy Agency
IAFS	International Affairs Programme (IAFS)
IARC(s)	International Agricultural Research Centre(s).
IARI	Indian Agricultural Research Institute, New Delhi.
IASRI	Indian Agricultural Statistics Research Institute
IBF	India business forum
IBGE	Institute Brasiliero de Geografia e Estadistica
IBPGR	International Board for Plant Genetic Resources
IBRD	International Bank for Reconstruction and Development
IBRD	International Bank for Reconstruction and Development (The World Bank)
IBSA	India, Brazil and South Africa
IBSNAT	International Benchmark Soils Network for Agro-technology Transfer
ICAO	International Civil Aviation Organization
ICAR	Indian Council of Agricultural Research
ICARDA	International Center for Agricultural Research in the Dry Areas
ICDP	Intensive cattle Development project
ICITV	Institute de la Carte Internationale du Tpis vegetable
ICMR	Indian Council of Medical Research
ICOC	Indian Central Oilseed Committee
ICRAF	International Council for Research in Agro-forestry
ICRISAT	International Crops Research Institute for the Semiarid Tropics

Abbreviation	Full Form
ICSI	Institute of Company Secretaries of India
ICSU	International Council of Scientific Unions
ICT	Information and Communication Technology
ICT	Information Communication Technology
ICTC	Integrated Counseling and Testing Centre
ICU	Intensive Care Unit
IDA	International Development Association
IDDA	Integrated Development of Dry Areas
IdoA	Iduronic acid
IDP	Internally displaced production
IDRC	International Development Research Centre
IECWSL	ICRISAT Employees Cooperative Welfare Society
IEF	Isoelectric focusing
IFAD	International Fund for Agricultural Development
IFC	International Finance Corporation
IFDC	International Fertilizer Development Centre
IFM	Integrated Fertility Management
IFPRI	International Food Policy Research Institute
IGBP	International Goesphere Biosphere Programme
IGES	Institute for Global Environment Strategies
IGFRI	Indian Grassland and Fodder Research Institute
IIAER	Indian Institute of Agricultural Economics Research
IIASA	International Institute for Applied Systems Analysis
IIHR	Indian Institute of Horticultural Research
IIMI	International Irrigation Management Institute
IIP	Index for Industrial production
IIPR	Indian Institute of Pulses Research
IISR	Indian Institute of Sugarcane Research
IISS	Indian Institute of Soil Science
IIT	Indian Institute of Technology

Abbreviation	Full Form
IITA	International Institute of Tropical Agriculture
IJO	International Jute Organization
ILCA	International Livestock Centre for Africa
ILO	International Labour Organization
ILRAO	International Laboratory for Research on Animal Diseases
ILRI	Indian Lac Research Institute, Ranchi.
ILRI	International Livestock Research
IMAGE	Integrated model to assess the green house effect
IMCO	Inter-Governmental Maritime Consultative Organization
IMF	International Monetary Fund
IMMA	Indian Micro-Fertilize Manufacture Association
IMY	Indira Mahila Yojana (1995)
INCOIS	Indian National Center for Ocean Information Services
INMS	Integrated Nutrient Management System
INRA	L' Institute Nacionale de la Recherche agronomique
INSFFER	International Network on Soil Fertility 'and Fertilizer Evaluation on Rice (IRRI)
INSIMP	Initiative for Nutritional Security through Intensive Milles Promotion
IONS	Indian ocean naval symposium
IOSM	International organization of securities commission
IPCC	Inter government penal on climate change
IPM	Integrated Pest Management
IPR	Intellectual Property Right
IQ	Intelligence quotient
IRDA	Insurance Regulatory and Development Authority
IRDC	International Research and Development Council
IRDP	Integrated Rural Development Programme
IRMA	Indian Research & Management Academy
IRRI	International Rice Research Institute
IRS	Infrared spectroscopy

Abbreviation	Full Form
IRTP	International Rice Testing Programme
ISBN	International Series of Book Number
ISH	International School of Hyderabad
ISI	Indian Statistical Institute
ISNAR	International Service for National Agricultural Research
ISRIC	International Soils Reference and Information Centre
ISSN	International Series Subscribe/Scientific Number
ISSS	International Society of Soil Science
ITC	Information Technology Center
ITK	Indigenous Technological Knowledge
ITO	International Trade Organization
ITU	International Telecommunication Union
IUCN	International Union for Conservation of Nature and Natural and Natural Resources
IUFRO	International Union of Forest Research Organizations
IVLP	Institute Village Linkage Programme
IVRI	Indian Veterinary Research Institute
IWM	Integrated Water Management
IWMI	International Water Management Institute
J	Joule
J&K	Jammu and Kashmir
JARI	Jute Agricultural Research Institute
JC	College of Agriculture
JDA	Joint Director of Agriculture
JECFA	Joint expert committee of food additives
JGSY	Jawahar Gram Samrudhi Yojna
JICA	Japan international cooperation agency
JKUAST	Sher-e-Kashmir Jammu and Kashmir University of Agricultural Sciences and Technology
JMDC	Jute manufactures development council
JNKVV	Jawaharlal Nehru Krishi Vishwa Vidyalaya

Abbreviation	*Full Form*
JRF	Junior Research Fellow
JRY	Jawahar Rozgar Yojna (1989)
JTRL	Jute Technological Research Laboratory
K	Kelvin
K	Kilo
KAU	Kerala Agricultural University
Kcal	Kilocalorie
KDa	Kilodalton
KJ	Kilojoule
KKV	Konkan Krishi Vidyapeeth
KMS	Potassium metabi sulphite
KPa	Kiloascal
KSI	Knowledge Sharing and Innovation
KVK	Krishi Vigyan Kendra
KYC	Know Your Customer
L	Liter
LAN	Local area network
LBPV	Lower Brahmaputra Valley
LCD	Liquid Crystal Display
LCIA	London court of international arbitration
LLP	Lab-to-Land Programme
LR	Lime Requirement
Ltd	Limited
LU	Lucknow University
LUC	Land-use category
LULUCF	Land use, Land-use change and forestry
M	mega
m	meter
M	Molar
M/s	Messrs

Abbreviation	Full Form
MACS	Maharashtra Association for Cultivation of Sciences
MAEER'S	Maharashtra Academy of Engineering & Educational Research
MAIDITOF	Matrix assisted laser desorption/ionization
Man	Mannose
MANAGE	National Institute of Agricultural Extension Management
MAU	Marathwada Agricultural University
MBA or M.B.A.	Master of Business Adminstration
MBP	Maltose binding protein
MCP	Multiple Cropping Programme
MDRT	Million Dollar Round Table
MEA	Ministry of External Affairs
MFALA	Marginal Farmers and Agricultural Labours Agency
MFP	Manufacturing Food Price
mg	milligram
MGREGA	Mahatma Gandhi Rural Employment Guarantee Act
MIDDA	Model for Innovation Development Diffusion & Adoption
min	minute
MIP	Markets, Institutions and Polices
MIT	Massachusetts Institute of Technology or Maharashtra Institute of Technology.
MKP	Minikit Programme
ml	milliliter
MLPA	Multiplex ligation dependent pro-amplification
mm	millimeter
mmHg	millimeter of mercury (Pressure)
MNP	Minimum Needs Programme
MOFPI	Ministry of Food Processing Industry
mol	mole
MOU	Memo of understanding
MOU	Memorandum of Agreement

Abbreviation	Full Form
MPKV	Mahatma Phule Krishi Vidyapeeth
MPSC	Maharashtra Public Service Commission
MPSS	Massively parallel signature sequencing
Mr	Mister
MRP	Maximum Retail Price/Manufacturing Resource Planning
Mrs	Mistress
MRV	Measurable reportable and verifiable
MSC	Military Staff Committee
MSG	Mono sodium glutamate
MSI	Moisture sorption isothern
MSP	Minimum support price
MSY	Mahila Samriddhi Yojana
MTD	Mahanagr Telecommunication Nigam Limited
MUM	Maximal unique match
Mur	Muramic acid
Mur2Ac	N-Acetylmuramic acid
mV	millivolt
MVC	College of Veterinary Science, Madras
n	nano
N	Normal (concentration)
NAARM	National Academy of Agricultural Research Management
NABARD	National Bank for Agriculture and Rural Development
NACMCF	National Advisory committee of microbiological criteria for food
NACMFI	National Agricultural Co-operative and Marketing Federation of India
NAD	Nicotinomide adenine dinucleotide, energy precursor.
NADA	National anti-doping agency
NAEP	National Agricultural Extension Project
NAIP	National Agricultural Innovation Project
NAIS	National Agricultural Insurance Scheme

Abbreviation	Full Form
NAMA	Non-Agricultural market access
NAPCC	National action plan on climate change
NAREGA	National Rural Employment Gurantee Act or Mahatma Gandhi. National rural employment gurantee Act.
NARI	National Aids Research Institute.
NARP	National Agricultural Research Project
NATP	National Agricultural Technology Project
NBAGR	National Bureau of Animal Genetic Resources
NBAIM	National Bureau on Agriculturally Importance Microorganism
NBFGR	National Bureau of Fish Genetic Resources
NBFGR	National Bureau of Fish Genetic Resources
NBP Zone	North Bank Plains Zone
NBPGR	National Bureau of Plant Genetic Resources
NBRI	National Botanical Research Institute
NBSSLUP	National Bureau of Soil Survey and Land-Use Planning
NBTC	National Business and Technology Conference
NCBI	National Center for Biotechnology Information
NCCS	National Cell Council of Science
NCDEX	National Commodity and Derivatives Exchange Limited
NCHRH	National Council for Human Resource in Health
NCL	National Chemical Laboratory
NCPAH	National Committee on Plasticulture Application in Horticulture
NCPCR	National commission for protection of child rights
NCTC	National Control Terrorist Center
NDA	National Defence Academy
NDMA	National Disaster Management Academy
NDRI	National Dairy Research Institute
NDUAT	Narendra Dev University of Agriculture and Technology
NEERI	National Environmental Engineering Research Institute
NEH-RC	North-Eastern Hills Research Center

Abbreviation	Full Form
NELP	New exploration and licensing policy
NEPA	National environmental policy act, 1969
NESP	National Extension Service Programme
NET	National Eligiblity Test
Neu5Ac	N-Acetylneuramic acid (a Sialic acid)
NFBSFARA	National Fund for Basic, Strategic and Frontier Application Research in Agriculture
NFE	Nitrogen Free Extract
NFSM	National Food Security Mission
NFZ	No fire zone
NGGIP	National greenhouse gas inventories programme
NGO	Non-Government Organizations
NHM	National History Museum
NHM	National Horticulture Mission
NHRD	National Human Resource Development
NIA	National Investigation Agency
NIAG	National Institute of Animal Genetics
NIL	Normal inbreeding lines
NIN	National Institute of Nutrition.
NIV	National Virus Science Institute
nm	nanometer
No	Number
NOAEL	No observable adverse effect level
NODP	National Oilseed Development Programme
NOTA	None of the Above
NPK	Nitrogen, Phosphorus, Potash
NRCC	National Research Center for Camel
NRCE	National Research Center for Equines
NRCG	National Research Center for Groundnut
NRCM	National Research Center for Mushroom

Abbreviation	Full Form
NRCWA	National Research Centre for Women in Agricultural
NRCY&M	National Research Center for Yak and Mithun
NRDC	National Research Development Corporation
NREP	National Rural Employment Programme
NRHM	National Rural Health Mission
NSC	National Seed Corporation of India
NSS	National Service Scheme
NSSO	National Sample Survey Organization
NTCCO	National Training and Communication Center for Oilseeds
NWDPRA	National Watershed Development Programme for Rainfed Area
NYK	Nehru Yuva Kendra
OCR	Optical Character Recognition
OFT	On Farm Testing
OHP	Over Head Projector
OMR	Optical Mark Recognition
OPCW	Organization for the Prohibition of Chemical Weapons
OPTP	Oilseed Production Thrust Programme
ORP	Operational Research Project
ORW	Outline of Research work
OS	Oueme superieur
OSD	Officer on Special Duty
OU	Orissa University
OUAT	Orissa University of Agriculture and Technology
p	pico
P.M.	Time from Noon 12 to Midnight 12
P.S.	Post Script
P.T.O.	Please Turn Over
Pa	Pascal
PABA	p-aminobenzoic acid
PAC	Public accounts committee

Abbreviation	Full Form
PAGE	Pilot analysis of global ecosystems
PAN	peroxy acetyl nitrate
PAR	Photo-synthetically active radiations
PAU	Punjab Agricultural University
PC	Polyvinyl chloride or Project coordinator
PCR	Polymer chain reactions
PCRA	Petroleum Conservation Research Association
PDF	Probability distribution function/Portable Document Format
PDPP	Prevention of damage to public property
PDRT	Puen Division Round Table
PEI	Poverty environment institute
PEM	Protein-energy malnutrition
PFC	Partnership for change
PG	Post Graduate
PGFTR	Programme of management of forests and riparian land
Ph.D.	Doctor of Philosophy
PISA	Protein *in-situ* array
PITC	Phenyl iso-thio cynate
PKV	Punjabrao Krishi Vidyapeeth
PLA	Participatory Learning Appraisal
PMF	Peptide mass fingerprinting
PMRY	Prime Minster Rozgar Yojna
PMS	Peptide mass searching
PO	Probationary Officer
POSDCORB	Planning, Organizing, Staffing, Directing, Co-ordinating Reporting Budgeting
PPM	Part Per Million
PPTD	Pilot Project for Tribal Development
PPW	Prepared plan of work
PRA	Participatory Rural Appraisal

Abbreviation	Full Form
PRII	Potash Research Institute of India
Pro	Proprietor
Prof	Professor
PSDS	Purchase, Supplies and Disposal Services
PT	Polyethylene terephatlate
PTD	Participatory Technology Development
PU	Punjab University
PVD	Physical vapour deposition
PVP	Plant Variety Protection
QA/QC	Quality Assurance/Quality control
QTL	Quantitative trait loci
r	revolution
RAM	Random access memory
RAPD	Random amplified polymorphic DNA
RAU	Rajendra Agricultural University
RBD	Random Block Design
RBS	Raja Balwant Singh College, Bichpuri
RDA	Recommended dietary allwance
RDS	Resilient Dryland Systems
REC	Regional Extension Centre
Ref	Reference
RFLP	Restriction fragment length polymorphism
RFRS	Regional Fruit Research Station
RH	Relative Humidty
Rha	Rhamnose
Rib	Ribose
RIL	Restricted inbreeding lines
RKVY	Rashtria Krishi Vikas Yojana
RLEGP	Rural Landless Employment Guarantee Programme
Rly	Railway

Abbreviation	Full Form
RMK	Rashtriya Mahila Kosh
RML	Royal market light
RPM	Resource Planning and Marketing
RRA	Rapid Rural Appraisal
RRS	Regional Research Station
Rs	Rupees
RSNC	Royal Society for Nature Conservation
RTI	Right to information
RV	Real Value
RWH	Rain Water Harvesting
S	Second
S	Svedberg unit
SAARC	South Asian Association for Regional Cooperation
SAGE	Serial analysis of gene expression
SAR	Second assessment report
SAR	Sodium Adsorption Ratio
SAR	Systematic acquired resistance
SAREC	Swedish Agency for Research Co-operation
SBI	Sugarcane Breeding Institute, Coimbatore
SC	Scheduled Caste
SCU	Sulphur Coated Urea
SDA	Specific dynamic action
SDEO	Sub-divisional Extension officer
SDRD	Survey Design and Research Division
SEC	Security
SELDI	Surface ehanced laser desorption/ionization
SET	State Eligible Test
SFDA	Small Farmer's Development Agency
SHG	Self- Help Groups
SIDA	Swedish International Development Agency

Abbreviation	*Full Form*
SITE	Satellite Instructional Television Experiment
SKNCA	Shri Karna Narendra College of Agriculture
SMS	Short Message Service
SMS	Subject Matter Specialists
SMSF	Suri M Sehgal Foundation
SNP	Single nucleotide repeats
SOC	Soil organic carbon
SPI	Soil Productivity Index
SPOT	System probatoire d' observation de la terre
SPR	Surface plasmin resonane
SPS	Sanitary and phytosanitary measures
SRES	Scientific research & assays
SRF	Senior Research Fellow
SRW	Synopsis of Research Work
SSLP	Simple sequence length polymorphism
ST	Scheduled Tribe
STEP	Support to training & Employment Programme
STMS	Sequenced taged microsatellite site
STR	Short tandom repeats
STSM	Sequence tagged site mapping
SWMA	Standard weight and measure act
TAD	Tribal Area Development Project
TANSACS	Tamil Nadu State Aids Control Society
TAR	Technology Assessment & Refinement
Tc	Collapse temperature
tC	Tonne carbon
TDI	Tolerable daily intake
tdm	Tonne dry matter
TDN	Total Digestible Nutrients
TDS	Total dissolved Solids

Abbreviation	Full Form
TDS/TCS	Tax Deduction Source/Tax collection Source
TE	Eutectic temperature
TLC	Thin Layer Chromatography
TLLING	Targeting induced local lesion in genome
TMO	Technology Mission on Oilseeds
TNAU	Tamil Nadu Agricultural University
TOF	Time of flight
TOGA	Total gene expression analysis
TOT	Transfer of technology
TPRI	Tropical Pesticides Research Institute
TQM	Total Quality Management
TRI	Telecommunication Regulatory Institute
TRIPS	Trade related aspect on intellectual property right
TRYSEM	Training Rural youth for Self Employment
TSU	Technical support unit
TTC	Trainer's Training Centre
TTK	Tchaouro-Toui-Kilibo
UAS	University of Agricultural Sciences
UC	Utilization Certificate
UG	Under Graduate
UIDAI	Unique identification authority of India
UMPP	Ultra mega power project
UNCTAD	United Nations Conference on Trade and Development
UNDFEW	United Nations Development Funds for Women
UNDOF	United Nations Disengagement Observer Force
UNDP	United Nations Development Programme
UNEP	United Nations Environment Programme
UNESCO	United Nations Educational, Scientific and Cultural Organization
UNFCCC	United nations framework convention on climate change
UNFICYP	United Nations Peace-Keeping Force in Cyprus

Abbreviation	Full Form
UNFPA	United Nations Fund for Population Activities
UNHCOR	United Nations High Commissioners Office for Refugees
UNICEF	United Nations Children Fund
UNIDO	United Nations Industrial Development Organization
UNIFIL	United Nations Interim Force in Lebanon
UNITAR	United Nations Institute for Training and Research
UNMOGIP	United Nations Military Observer Group in India and Pakistan
UNRWA	United Nations Relief and Works Agency
UNSF	United Nations Special Fund
UNSSO	United Nations Sudana-Sahelian Office
UNTSO	United Nations Truce Supervision Organization
UNU	United Nations University
UPSC	Union Public Service Commission
UPU	Universal Postal Union
USAID	United States Agency for International Development
UU	University of Udaipur
UVB	Ultraviolet
V	Volt
VC	Veterinary College.
VCR	Video Cassette Recorder
VCTC	Voluntary Counseling and Testing Centers
VEW	Village Extension Worker
Viz	Namely
VNTR	Variable number of tandom repeats
VPKAS	Vivekananda Parvatiya Krishi Anusandhan Shala
WALMI	Water & Land Management Institute
WARDA	West African Rice Development Association
WCED	World commission on environment and development
WCLA	Economic Commission for Latin America
WFC	World Food Council

Abbreviation	*Full Form*
WFP	World food programme
WHO	World Health Organization
WIPO	World Intellectual Property Organization
Wm	monolayer water content
WMO	World Meteorological Organization
WRB	World reference base for soil resources
WRI	World resources institute
WSO	Without Sandalwood Oil
WTSS	Whole transcriptions short gun sequencing
WUE	Water use efficiency
WWF	World wide fund
Xyl	Xylose
Yr	Year
ZEO	Zonal Extension Officer
ZSI	Zoological Survey of India, Calcutta

3

PRINCIPLES

➤ **Principle of Extension Education**

1. Principle of cultural difference
2. Grass roots Principle
3. Principle of indigenous knowledge
4. Principle of interests & needs.
5. Principle Participation
6. Principle Participation
7. Principle whole Family approach
8. Principle leadership
9. Principle adaptability
10. Principle Satisfaction
11. Principle Evaluation

➤ **Principles of Learning**

1. Principle of self-activity
2. Principle of association
3. Principle transfer
4. Principle of disassociation
5. Principle readiness
6. Principle attitude
7. Principle Practice
8. Principle motivation
9. Principle timing
10. Principle Clarity of objectives
11. Principle Satisfying.

➤ **What are the criteria for effective extension teaching?**

1. Requires Specific & clearly defined teaching objectives.

2. Requires a Suitable learning Situation

3. Requires effective communication

4. Requires both content & method

5. Extension teaching must be looked upon as an intentional

6. Extension teaching must result in effective learning

7. Extension teaching must accomplish certain kinds of educational changes in relation to the subject matter taught.

8. Extension teaching requires careful evaluation Extension.

> **What are the principles of extension programme planning?**

1. Extension Programme should be based on an analysis of the past experiences, Present situation & future needs.

2. Extension Programmes should have clear & significant objectives which could satisfy important needs of the people.

3. Extension Programmes should fix up priority on the basis of available resource resources & time.

4. Extension Programmes should have a general agreement at various levels.

5. Extension Programmes should involve relevant institutions & organizations.

6. Extension Programmes should involve people at the local level.

7. Extension Programmes should involve relevant institutions & organizations.

8. Extension Programmes should have definite plan of work.

9. Extension Programmes should provide for evaluation of results & reconsideration of the programme.

10. Extension Programmes should provide for equitable distribution of benefits amongst the members of the community.

> **What are the steps in extension programme planning?**

1. Collections of facts

2. Analysis of situation

3. Identification of problems

4. Determination of objectives & goals

5. Developing plan of work & calendar of operations

6. Follow through plan of work & calendar of operations.

7. Evaluation of progress.

8. Reconsideration & revision of the programme.

➢ **Models of Communication**

1. Aristotle model (384-322 B.C)

    Speaker → Speech → Audience

2. Sharon-weaver model (1949)

    Source → Transmitter → Signal → Receiver

3. Berlo model (1960)

    Source → Encoder → Message → Channel

4. Schramm Model

    Source → Encoder → signal → Decoder → Destination

5. Lagans Mode (1963)

    Communicator → Message → Channel → Treatment →

    Audience → Response

6. Rogers & Shoemaker Model (1979)

    Source → Message → Channel → Receiver → Effects

A Source (S) Sends a message (M) via certain channels (C) to the receiving individual (e) which causes some effects (E) i.e. changing the existing of the receiver.

EXAMPLES OF COMMUNICATION METHODS

Individual

1.	Personal contact	4.	Telephone Call
2.	Letters	5.	Result demonstration
3.	Office Call	6.	Farm & home visit

Group Contact method (2-30 Persons)

1.	Discussion	6.	Farm School
2.	Lecture	7.	Meeting
3.	Conference	8.	Field day
4.	Seminar	9.	Method/Result demonstration
5.	Training Camps	10.	Flash card

Mass or Community Contact Method

(1) Bulletins   (2) News Papers   (3) Journal

(4) Magazine   (5) Exhibition   (6) Circular letter

(7) Fairs   (8) Posters   (9) Cinema (Film)

(10) Movie   (11) Drama and songs   (12) Radio

(13) Leaflet/Folder   (14) Television

CLASSIFICATION OF AUDIO–VISUAL AIDS

Audio aids	Visual aids Non–projected	Audio Visual aids Non–projected
1. Tape recorder	1. Chalk board	1. Drama, puppet,
2. Public address system	2. Bulletin board	Show talking
3. Telephone	3. Picture and Photograph doll	
	4. Flannel graph, flash card, flip char	
	5. Poster	
	6. Diagram, map, chart,	
	And graph	
	7. Specimen, model, diorama	
	8. Translate	
	Projected	**Projected**
	1. Slides	1. Motion picture (Cinema)
	2. Filmstrip	2. Video
	3. Opaque projection	
	4. Overhead projection	

4

VIEWPOINTS

- Rural participatory appraisal is related with agriculture extension.
- Success of rural projects depends upon partnership of beneficiaries.
- First time the word "extension" was used in USA.
- Gurgaon attempt was organized by F.L.Bryne, collector of Gurgaon district in Haryana in 1920.
- The comilla project approach was first used by the academy of rural development at comilla in Bangladesh.
- The principle "One village one society" was suggested by Meelogen committee.
- The rural System research idea was motivated by M.S. Swaminathan in 1988.
- Selecting "Poorest among poor first" is the approach under integrated rural Development programme.
- Albert Mayer is the name associated with Etawah Pilot project.
- The first Krishi Vigyan Kendra in India was established in Pondicherry.
- A round table discussion is called panel.
- ATMA (Agricultural Technol.ogy Management Agency) is a registered society associated with IVLP (Institutional Village Linkage Programme)
- Total number of KVK operating in the country is now 540 & in Maharashtra is 33
- The basic unit of development under I.R.D. programme is family.
- Extension workers should be verse with economic principles without it they cannot give effective advice form business point of view.
- The commission recommending the minimum support prices for farmers is known as commission for agriculture costs and prices.
- In India, community development projects were launched in 1952.
- Audio-visual aid is television.
- The T and V system was proposed by Dr. Danial Benor.

- TRYSEM scheme was started in 15110 Aug. 1979.
- Integrated Rural Development Programme was started in 2nd Oct. 1980.
- The extension education aim at changing knowledge, skills and attitudes.
- Extension Education is different from formal education in a sense that we start from practical's and end up with theory.
- IVLP stands for Institutions village linkage programme.
- Krishi Vigyan Kendra's (KVKSs) can only under control of agricultural university and voluntary organization.
- The "Lab to Land" programme also included demonstration but they were different from early demonstrations in terms of size of demonstration.
- District credit plan is prepared by Lead Bank.
- "Operation flood" programme relates to boosting mill supply.
- Television fairs. Radio and exhibition are the media of mass communication.
- UPDASP is sponsored by world bank.
- Rural participatory appraisal is related with agriculture extension.
- High yielding variety programme was started in the year 1966.
- U.P.D. A.S.P. stands for U.P. diversified agriculture support project.
- The prime responsibility of the transfer of technology in Rajasthan State with agriculture department.
- Puppet is a traditional and simple aid of communication.
- August Compute is the father of Sociology.
- Scope of sociology is studied community.
- Panchayati Raj in India was first recommended by Balwant Rai Mehta committee. National Extension services was started in the year 1953.
- The term "Human communication" means exchange of ideas.
- The term "inter-communication" means communication on any issue, communication between to persons, communication between various groups.
- Evaluation done by the agency is known as external evaluation.
- In extension education exhibition is helpful for mass communication.
- High yielding variety programme has given emphasis on package practice with dwarf varieties.

- The method for establishing "mass contracts" is Bulletin.
- Drama is audio visual aid.
- Easterner Agricultural University is located in Imphal
- Extension education is an educational process.
- Knowledge, skill attitude is the element of behaviour.
- Family is included under primary group.
- Nilokheri experiment was started by Albert Mayer.
- The process of initiating a conscious and purposeful action is known as motivation.
- "A listing of activities by which the objectives already decide" up to shall be achieved this definition of plan of work was given by A. H. Maunder.
- Kisan Mandai System under T and V system was initiated in the year of 1993 Rural Development programme at district level is being looked after by DRDA.
- In extension education communication is a expression of thoughts.
- Method Demonstration is the most effective approach of the extension of the group of the people.
- Etowah pilot project was first time introduced in the year of 1952.
- The Firka development project was first time introduced in the state of Tamil Nadu.
- First KVK was established in India in the year of 1974. In the Pondicherry. Training and visit progrmme was first time introduced in Rajasthan and M.P.
- The word 'tension' in extension is derived form Latian word.
- J. Paul Lagans is a father of extension education.
- First Model of communication was given by Berlo.
- Panchayati Raj report was submitted by Balwant Rai Mehta in the year of 1957.
- First KVK established in Rajasthan was Fatehpur, 1976
- F.L. Brayne was started the Gurgaon experiment.
- "Jawahar Rozgar Yojana" was introduced in April 1989.
- National Agriculture Research project was launched by ICAR for location specific research in the year 15 Jan. 1979.
- T and V system was introduced with the assistance of World Bank.

- Socio-metric method is best suited for selection of leader.
- "Grow more food" programme was started in India in the year of 1945.
- The main purpose of T and V programme is to help farmers in the adoption of latest innovation in agriculture.
- An extension worker practically explains Japanese method of paddy cultivation to a group of farmers and also shown its superiority over the traditional method. In the process method and result demonstration communication method have been used by him.
- First KVK established in Maharashtra was Koshbad at Thane

IMPORTANT TIPS

1. Sociology is the science of society.
2. Trait is mode of behaviour.
3. Democratic decentralization was recommended by R.R. Mehta.
4. The fundamental objective of community development is overall development of the people.
5. Extension is helping the people to help themselves.
6. The goal of teaching is to bring about the desired change in the behaviour of the "Audience".
7. Social class is an open class system whereas caste is the closed class system.
8. The term "Sociology" was coined by August Compute who is often revered to as the father of sociology.
9. Social structure is composed of group.
10. Organizations, institutions and community are referred to as forms of Human Association.
11. Group is a unit of two or more people in interaction and communication with each other.
12. KVK was recommended by Dr. Mohansing Mehata Committee.
13. The technique of important training to the youth club members is based on the principle of learning by doing and earning while learning.
14. Motivation is known as goal directed or goal seeking behaviour.
15. The state of being moved or stirred up feeling in one way or the other is called motion.

ROLE OF EXTENSION EDUCATION IN AGRICULTURAL DEVELOPMENT

1. Making the adoption of new technology possible.

2. To help the farmer to accept new agricultural technology for obtaining higher yields.

3. Extension workers' carry out field demonstrations about new technology.

4. Helps the farmers to visualize farming as a profitable occupation.

5. Helps farmers to understand their own situation, resources and critical analysis helps them to take rational decisions for the best use of the available resources.

6. Trains the local leaders.

7. Helps getting necessary inputs.

8. Helps in decision-making.

9. Helps to change attitude.

10. To bring about this change through education.

11. It helps in achieving cooperation of people.

12. It develops, strengthens and organizes groups, institutes and people to achieve their aims.

13. Extension helps the planners, policy-makers and administrators with local conditions and latest technologies suited to it.

14. Job of extension is to bridge the gap between the scientists on the one hand and the farmer's on the other. This is done by taking new farm technology and innovations to the farmers' door and helping him to adopt them.

It also brings the farmer's problems to the research stations for further investigations and to find out the required solutions.

Training programmes for extension workers : The process of training employees to gain effectiveness in their present or future work through the development of appropriate habits of thought and action, skills, knowledge and attitude knowledge.

IMPORTANCE OF TRAINING

1. Communication of the desired knowledge.

2. Knowledge of the techniques, job efficiency.

3. Syllabus of trainee should be according to the needs of extension agents.

4. Teaching of manual skills.

TYPE OF TRAINING

1. Pre-service training

2. In-service training

STEPS FOR TRAINING

1. Oriented towards the organizational setup.

2. Induction training is given at the orientation centres where the extension workers are given the knowledge of working and organization of community development and extension service.

3. Short-range courses organized by the state department of agriculture through state agriculture universities.

4. Periodical meetings and conferences are organized by state agriculture universities in collaboration with the state agriculture department for extension officers at district level or above.

MERITS

1. Training improves a person's skill and develops in him the desired attitude and values required for his work.

2. Training helps to acquire occupational work skills and the latest knowledge.

3. Makes him familiar with the objectives of the organization. It makes this contribution in promoting the goals of his organization.

LAB TO LAND PROGRAMME

The Lab-to-Land Programme (LLP) was launched by the ICAR in 1979 as a part of its Golden Jubilee celebration. The overall objective of the programme was to improve the economic condition of the small and marginal farmers and landless agricultural labourers, particularly scheduled castes and scheduled tribes, by transfer of improved technology developed by the agricultural universities, research institutes etc. The specific objectives of the lab to-land programme.

1. Study and understand the background and resources of the selected farmers and landless agricultural labourers. To introduce low-cost relevant agricultural and allied technologies on their farms and homes for increasing their employment, production and income.

2. Assist the farmers to develop feasible farm plans keeping in view the availability of technologies, needs and resources of the farmers, and the resources which could be made available from external sources and agencies.

3. Guide and help the farmers in adopting improved technologies as per their farm plans, and demonstrate to them the economic viability of those technologies as well as methods of cultivation and farm management.

4. Organize training programmes and other extension activities, in relation to their adopted practices, and prepare them for active participation in agricultural development programmes of the State.

5. Make the farmers aware of the various opportunities and agencies which they could utilize to their economic advantage.

6. Develop functional relations and linkages with the scientists and institutions for future guidance, advisory services and help.

7. Utilize this project as a feedback and mechanism for the agricultural scientists and extension functionaries.

The programme was initiated with 75,000 farm families over the whole country. Major thrust of the programme was to introduce the most and appropriate technologies that would help in the diversification of labour use and introduction of supplementary sources of income. The programme had been in operation in a number of phases. Normally a phase ran for a period of two years with a particular set of farmers.

TRAINING AND VISIT SYSTEM

By the middle of 1970s it was felt that extension services in the developing countries were suffering from a number of weaknesses, including the dissipation of extension workers' energies on low priority tasks; the lack of a single, clear line of command; and a low level of agricultural knowledge and expertise among field level functionaries. As a means of reforming and strengthening the extension services, a reorganized system of agricultural extension, known as Training and Visit (T&V) system was introduced in India in 1974 with the World Bank assistance. It was presumed that transfer of technology through the 'contact farmers' shall benefit all farmers.

In two and a half decades, T&V became the dominant method of restructuring the extension services in over sixty countries in Asia, Africa and Latin America. The system tries to achieve changes in production technologies used by the majority of farmers through assistance from well-trained extension agents who have close links with agricultural research and supported by supply, service and marketing facilities.

T&V was regarded as an improved management system of agricultural extension and had the following *key features* :

1. *Professionalism:* Each extension agent is fully and continuously trained to handle one's particular responsibilities in a professional manner.

2. *Single line* of *command:* The extension service must be under a single line of technical and administrative command within the Ministry/ Department of Agriculture.

3. *Concentration* of *effort:* All extension staff works only on agricultural extension. They are not responsible for any other activity not directly related to extension. In training sessions, attention is concentrated on important major points.

4. *Time-bound work:* Messages and skills are taught to farmers in a regular and timely fashion. The Village Extension Worker (VEW) must visit the farmers regularly on a fixed day, usually once each fortnight. All other extension staff must make timely and regular visits to the field. Recommendations for a specific area and for particular farming conditions for each two-fortnight periods are discussed and learned by Subject Matter Specialists (SMSs) at regular monthly workshops; the recommendations are then presented to YEWs and Agricultural Extension Officers.

5. *Field and farmer orientation:* The contact with the farmers must be on a regular basis, on a schedule known to farmers, and with a large number of farmers representing all major farming and socio-economic types.

6. *Regular and continuous training:* Regular and continuous training of extension staff and discuss with them, and to prepare specific production recommendations required by farmers for the coming fortnights and to upgrade and update their professional skills.

    Fortnightly training and monthly workshops are the key means of bringing actual farmers' problems to the attention of research, of identifying research findings of immediate relevance to farmers.

7. *Linkages with research:* Problems faced by farmers that cannot be resolved by extension agents are passed on to researchers for either an immediate solution or investigation. Seasonal and monthly workshops, joint field visits, training of extension staff and formulation of production recommendations are some of the means by which linkages with research are maintained.

ORGANIZATION OF T&V

The T and V system has a hierarchical organization with one extension agent (VEW) for about 800 to 1000 farmers. The farmers are divided into eight groups of about equal size. Ten farmers in each group are selected as *'contact'* farmers. Extension message is to percolate to other farmers through these contact farmers. Eight YEWs are supervised by an AEO, and eight AEOs are supervised by a Sub-Divisional/District Extension Officer (SDEO/DEO), who is assisted by three to five Subject Matter Specialists (SMSs). Higher level supervision is provided by Zone Extension Officer (ZEO) and Director of Extension/Director of Agriculture of the State, with the assistance of SMSs.

OTHERS

Socio-economic survey and status of big, small and marginal farmers and landless agricultural labourers.

INTRODUCTION

When planning rural development of a country, improving the socio-economic condition of the rural population is important.

In an agricultural-dominating economy, the landless labourers, marginal and small farmers are treated as target group, who should receive a high order of priority in the strategy of planning for rural development.

AGRICULTURE IS THE DOMINANT SECTOR OF THE INDIAN ECONOMY

1. GDP contribution in 2.5%.

2. Arable land is 143 in ha.

3. Farm holding is about 89 million.

4. Average size of holding is 1.84 ha.

5. The small and marginal farmers' holdings constitute around 75% of the total holdings in our country.

6. Small farmers have increased due to subdivision of medium-sized holdings by their heirs.

7. Marginal farmers have also grown in number when small farmers are shared by the heir.

8. Large holdings have become medium-sized holdings due to the operation of land ceiling act under land reforms.

9. According to National sample survey, 63.0% of the holdings are < 2 ha and they occupy nearly 19% of the total cultivated land.

10. Agricultural work force of 299 million in India (1931) which has grown to nearly 74% of this work force is agricultural.

The landless labourer has no self-employment; they depend entirely on daily wages for their living. This category, constitutes, the core of the employment problem in the rural sector.

CHARACTERISTICS

Small Farmers and Marginal Farmers

1. Poor land use

2. Surplus family labour

3. Under nutrition

4. Higher debt

5. Distress sale of farm produce

ECONOMIC CONDITION

Condition	Small farmers	Landless labourers
Work force	65.31%	55.72%
Household engaged in agriculture	91.05%	98.90%
Literacy	37.45%	26.50%
Bullock power	13.43%	11.11%
Own bullock carts	2.2%	2.2%
Average gross income	Rs. 14,000	Rs. 3,400
Per capita income	Rs. 2,600	Rs. 765
An average consumption	Rs. 894	Rs. 623

SOCIAL STATUS

The status of small and marginal farmers varies only marginally and that of landless agricultural labourers is slightly less than the above 2 categories.

EXTENT OF SUB-DIVISION AND FRAGMENTATION

Problems of small holdings

1. Small size holdings are many.

2. Many owners have small pieces of plots, so there is difficult to cultivate them together in an efficient manner.

Lack of capital surplus for investment

Majority have less area (1 ha), so they are cultivated and get only less yield, so this is insufficient for them to create capital surplus investment in land development.

Status of big farmers

1. They can mobilize surplus capital for investment.

2. Land development.

3. More land under cultivation.

4. Using modern equipment.

5. Green revolution benefited the big farmers more.

6. Than marginal and small farmers.

FARM MECHANIZATION

It is the application of engineering and technology in agricultural operations to do a job in a better way to increase productivity. It is a need-based process which provides sufficient time gap for self-adjustment of various inputs without causing sudden impact of changes.

SCOPE

1. Increased irrigation facility.

2. Introduction of HYV and new crops.

3. Introduction on high dose of fertilizers and pesticides.

4. Multi-cropping system and intensive cultivation.

BENEFITS

1. Timeliness of farm operation

2. Precision of operation

3. Enhancement of work and safety environment

4. Reduction of less crops and food products

5. Increased productivity of land

6. Increased economic return to farmer

7. Progress and prosperity in rural areas

9. To perform difficult operations which cannot be done by animal and manpower

10. Proper utilization of inputs like water, seed, fertilizer etc.

LIMITATIONS

1. Small land holdings (1.55 ha)

2. Low investing capacity of farmers

3. Agriculture labour is easily available

4. Adequate availability of draft animals

5. Lack of suitable farm machine for different operations

6. Lack of trained manpower

7. Lack of co-ordination between research organization and manufacturer

8. Inadequate quality control.

9. Lack of repair and service facilities for machines.

FUTURE NEEDS OF FARM MECHANIZATION

1. To develop multifunctional machines.

2. Development of combination machinery like minimum tillage, *zero* tillage, till planter etc.

3. Better utilization of solar, wind, energy and bio-fuels.

4. Efficient fuel utilization -through better design and matching machinery.

5. Rain water harvesting and conservation.

6. Low cost micro irrigation and sprinklers using surface water needs to be developed and commercialized.

CLASSIFICATION OF VARIOUS EXTENSION TEACHING METHODS

According to the use and nature of contact:

Sr. No.	Individual contact	Group contact	Mass community contact
1.	Face to face or person to person contact between the rural and extension workers.	Group consists of 20-25 persons groups are formed around common interest.	Large numbers of people are approached at short time to disseminate new information/technology.
2.	It is very effective in teaching new technology or skills.	It involves face to face exchanges and discussions.	It is the quick method for mass communication.
3.	Examples :	Examples :	Examples :
	1. Farm and home visits	1. Method and results demonstrations	1. Bulletins
	2. Office calls	2. National demonstration	2. Leaflets
	3. Telephone calls	3. Leader-training meetings	3. Newspapers
	4. Personal letters	4. Conferances	4. Radio
		5. Discussion meetings	5. Circulars
		6. Workshops	6. Television
		7. Field trips	7. Posters
		8. Forums	8. Exhibitions
		9. Symposiums	9. Fairs
			10. Campaings

According to forms of communications

Sr. No.	Written	Spoken	Objective or visual
1.	Bulletins	General and special meetings	Result demonstration
2.	Leaflets	Farm and home visits	Demonstration by posters
3.	Folders	Official calls	Motion pictures or movies
4.	News articles	Telephone calls	Charts
5.	Personal letters	Radio	Slides
6.	Circulars		Film strips & models

GIVE THE GENERAL VIEW OF UNITED NATIONS SYSTEM

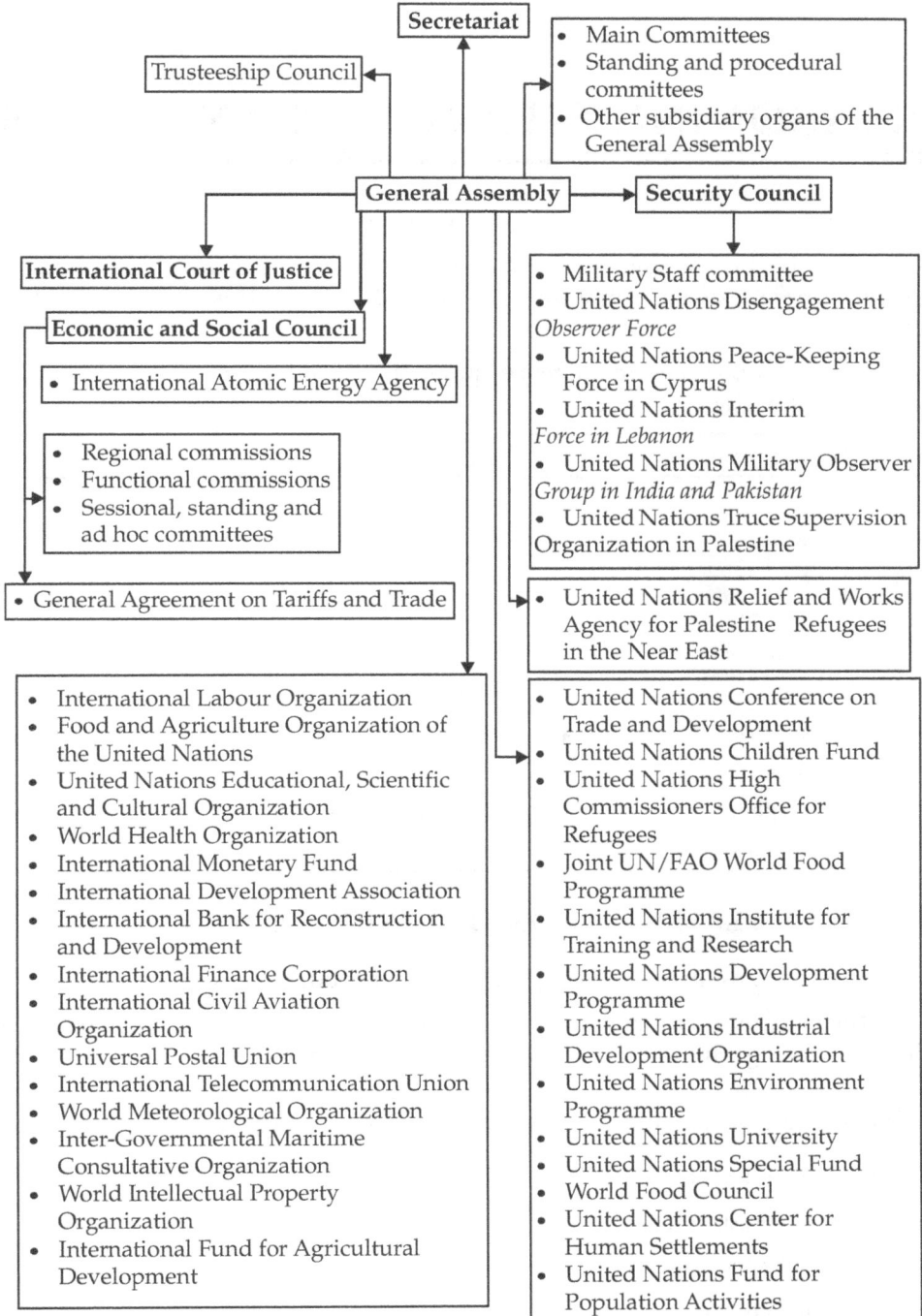

Secretariat

Trusteeship Council

- Main Committees
- Standing and procedural committees
- Other subsidiary organs of the General Assembly

General Assembly

Security Council

International Court of Justice

Economic and Social Council

- International Atomic Energy Agency

- Regional commissions
- Functional commissions
- Sessional, standing and ad hoc committees

- General Agreement on Tariffs and Trade

- International Labour Organization
- Food and Agriculture Organization of the United Nations
- United Nations Educational, Scientific and Cultural Organization
- World Health Organization
- International Monetary Fund
- International Development Association
- International Bank for Reconstruction and Development
- International Finance Corporation
- International Civil Aviation Organization
- Universal Postal Union
- International Telecommunication Union
- World Meteorological Organization
- Inter-Governmental Maritime Consultative Organization
- World Intellectual Property Organization
- International Fund for Agricultural Development

- Military Staff committee
- United Nations Disengagement *Observer Force*
- United Nations Peace-Keeping Force in Cyprus
- United Nations Interim *Force in Lebanon*
- United Nations Military Observer *Group in India and Pakistan*
- United Nations Truce Supervision Organization in Palestine

- United Nations Relief and Works Agency for Palestine Refugees in the Near East

- United Nations Conference on Trade and Development
- United Nations Children Fund
- United Nations High Commissioners Office for Refugees
- Joint UN/FAO World Food Programme
- United Nations Institute for Training and Research
- United Nations Development Programme
- United Nations Industrial Development Organization
- United Nations Environment Programme
- United Nations University
- United Nations Special Fund
- World Food Council
- United Nations Center for Human Settlements
- United Nations Fund for Population Activities

5

REASONING/SHORT EXPLANATIONS

What is intelligence test and how it is measured?

Intelligence tests measure to measure any person's ability to give answer quickly. These tests were first introduced by Dr. Binnet. It is also measured on the basis of I.Q. I. Q. = MA/CA x 100.

Where I.Q. means Intelligence quotient, MA means mental age and CA means calendar age of an individual.

S.No.	Level of intelligence	I.Q. Range
1.	Idiob	0 to 25
2.	Imbecide	26 to 50
3.	Moron	51 to 70
4.	Border line	71 to 80
5.	Below normal	81 to 90
6.	Normal	91 to 110
7.	Above normal	111 to 120
8.	Superior	121 to 130
9.	Very superior	131 to 140
10.	Near genius	Above 140

GIVE THE CLASSIFICATION OF VARIOUS GROUPS?

Primary group; Secondary group; Formal group; Informal group.

Social group: Horizontal group, Apex income group, Low income group, Small income group (30 units), Big group more than 30 units.

Give the classification of various types of villages?

1. Isolated village
2. Clustered village
3. Nucleated village
4. Line village

5. Round or circular village

6. Cross road villages

7. Hamlets villages (Indian villages)

8. Typical village

What are the duties of subject matter specialist?

1. Planning the programme

2. Training the staff

3. Teaching to the people

4. To make field study

5. Preparing teaching material to increase effectiveness of the work

6. To keep the village level worker up to date in their work

7. He serves as a bridge between the research department and village level worker

8. He collects and analyzes data and solved problems related to farmers field

9. He develops the community programme.

GIVE THE CLASSIFICATION OR GROUPS OF FAMILIES?

Family is basic institution or relationship between the parent and children.

1. **Nucleated family:** Joint family, grandfather, mother, parent, children, sisters, brothers, uncles, cousins live together.

2. **Separated family:** It is medium size family up to 5 persons live together.

3. **Extended family:** Husband, wife and child live together only.

According to residence

1. Matrilocal means husband goes to live with wife to her house.

2. Patrilocal means wife goes to live with husband at their house.

According to ancestors

1. Matrilinear means mother or wife is main ancestor.

2. Patrilinear means father or guardian is main ancestor.

According to authority

1. Matri channel means mother is the main authority in the house.
2. Patrichannel means father or guardian is the main authority in the house.

According to marriage

1. Monogamous means one man and one women or wife.
2. Polygamous means one man and more than one women or wife.
3. Polyandrous means one woman and more than man.

GROUP OF TWENTY COUNTRIES (G-20)

What is the G-20?

The Group of Twenty (G-20) Finance Ministers and Central Bank Governors was established in 1999 to bring together systemically important industrialized and developing economies to discuss key issues in the global economy. The inaugural meeting of the G-20 took place in Berlin, on December 15-16, 1999, hosted by German and Canadian finance ministers.

MANDATE

The G-20 is the premier forum for our international economic development that promotes open and constructive discussion between industrial and emerging-market countries on key issues related to global economic stability. By contributing to the strengthening of the international financial architecture and providing opportunities for dialogue on national policies, international co-operation, and international financial institutions, the G-20 helps to support growth and development across the globe.

ORIGINS

The G-20 was created as a response both to the financial crises of the late 1990s and to a growing recognition that key emerging-market countries were not adequately included in the core of global economic discussion and governance. Prior to the G-20 creation, similar groupings to promote dialogue and analysis had been established at the initiative of the G-7. The G-22 met at Washington D.C. in April and October 1998. Its aim was to involve non-G-7 countries in the resolution of global aspects of the financial crisis then affecting emerging-market countries. Two subsequent meetings comprising a larger group of participants (G-33) held in March and April 1999 discussed reforms of the global economy and the international financial system. The proposals made by the G-22 and the G-33 to reduce the world economy's susceptibility to crises showed the potential benefits of a regular international consultative forum embracing the emerging-market countries. Such a regular dialogue with a constant set of partners was institutionalized by the creation of the G-20 in 1999.

MEMBERSHIP

The G-20 is made up of the finance ministers and central bank governors of 19 countries:

- Argentina
- Australia
- Brazil
- Canada
- China
- France
- Germany
- India
- Indonesia
- Italy
- Japan
- Mexico
- Russia
- Saudi Arabia
- South Africa
- Republic of Korea
- Turkey
- United Kingdom
- United States of America

The European Union, who is represented by the rotating Council presidency and the European Central Bank, is the 20th member of the G-20. To ensure global economic fora and institutions work together, the Managing Director of the International Monetary Fund (IMF) and the President of the World Bank, plus the chairs of the International Monetary and Financial Committee and Development Committee of the IMF and World Bank, also participate in G-20 meetings on an ex-officio basis. The G-20 thus brings together important industrial and emerging-market countries from all regions of the world. Together, member countries represent around 90 per cent of global gross national product, 80 per cent of world trade (including EU intra-trade) as well as two-thirds of the world's population. The G-20's economic weight and broad membership gives it a high degree of legitimacy and influence over the management of the global economy and financial system.

ACHIEVEMENTS

The G-20 has progressed a range of issues since 1999, including agreement about policies for growth, reducing abuse of the financial system, dealing with financial crises and combating terrorist financing. The G-20 also aims to foster the adoption of internationally recognized standards through the example set by its members in areas such as the transparency of fiscal policy and combating money laundering and the financing of terrorism. In 2004, G-20 countries committed to new higher standards of transparency and exchange of information on tax matters. This aims to combat abuses of the financial system and illicit activities including tax evasion. The G-20 has also aimed to develop a common view among members on issues related to further development of the global economic and financial system.

To tackle the financial and economic crisis that spread across the globe in 2008, the G-20 members were called upon to further strengthen international cooperation. Accordingly, the G-20 Summits have been held in Washington in 2008, in London and Pittsburgh in 2009, and in Toronto and Seoul in 2010.

The concerted and decisive actions of the G-20, with its balanced membership of developed and developing countries helped the world deal effectively with the financial and economic crisis, and the G-20 has already delivered a number of significant and concrete outcomes:

First, the scope of financial regulation has been largely broadened, and prudential regulation and supervision have been strengthened. There was also great progress in policy coordination thanks to the creation of the framework for a strong, sustainable and balanced growth designed to enhance macroeconomic cooperation among the G-20 members and therefore to mitigate the impact of the crisis. Finally, global governance has dramatically improved to better take into consideration the role and the needs of emerging of developing countries, especially through the ambitious reforms of the governance of the IMF and the World Bank.

CHAIR

Unlike international institutions such as the Organization for Economic Co-operation and Development (OECD), IMF or World Bank, the G-20 (like the G-7) has no permanent staff of its own. The G-20 chair rotates between members, and is selected from a different regional grouping of countries each year. In 2011 the G-20 chair is France. The chair is part of a revolving three-member management Troika of past, present and future chairs. The incumbent chair establishes a temporary secretariat for the duration of its term, which coordinates the group's work and organizes its meetings. The role of the Troika is to ensure continuity in the G-20's work and management across host years.

FORMER G-20 CHAIRS

- 1999-2001 Canada
- 2002 India
- 2003 Mexico
- 2004 Germany
- 2005 China
- 2006 Australia
- 2007 South Africa
- 2008 Brazil
- 2009 United Kingdom
- 2010 Republic of Korea

MEETINGS AND ACTIVITIES

It is normal practice for the G-20 finance ministers and central bank governors to meet once a year. The ministers' and governors' meeting is usually preceded by two deputies' meetings and extensive technical work. This technical work takes the form of workshops, reports and case studies on specific subjects, that aim to provide ministers and governors with contemporary analysis and insights, to better inform their consideration of policy challenges and options.

INTERACTION WITH OTHER
INTERNATIONAL ORGANIZATIONS

The G-20 cooperates closely with various other major international organizations and for a, as the potential to develop common positions on complex issues among G-20 members can add political momentum to decision-making in other bodies. The participation of the President of the World Bank, the Managing Director of the IMF and the chairs of the International Monetary and Financial Committee and the Development Committee in the G-20 meetings ensures that the G-20 process is well integrated with the activities of the Bretton Woods

Institutions. The G-20 also works with, and encourages, other international groups and organizations, such as the Financial Stability Board and the Basel Committee on Banking Supervision, in progressing international and domestic economic policy reforms. In addition, experts from private-sector institutions and non-government organizations are invited to G-20 meetings on an ad hoc basis in order to exploit synergies in analyzing selected topics and avoid overlap.

EXTERNAL COMMUNICATION

The country currently chairing the G-20 posts details of the group's meetings and work program on a dedicated website. Although participation in the meetings is reserved for members, the public is informed about what was discussed and agreed immediately after the meeting of ministers and governors has ended. After each meeting of ministers and governors, the G-20 publishes a communiqué which records the agreements reached and measures outlined. Material on the forward work program is also made public.

1. **When was the G-20 set up?**

   The G-20 first meeting was held in Berlin on December 1516, 1999.

2. **Why was the G-20 set up?**

   The G-20 was created as a response both to the financial crises of the late 1990s and a growing recognition that key emerging-market countries were not adequately included in the core of global economic discussion and governance. Prior to the G-20 creation, similar groupings to promote dialogue and analysis had been established at the initiative of the G-7. The G-22 met at Washington D.C. in April and October 1998. Its aim was to involve non-G-7 countries in the resolution of global aspects of the financial crisis then affecting emerging-market countries. Two subsequent meetings comprising a larger group of participants (G-33) held in March and April 1999 discussed reforms of the global economy and the international financial system. The proposals made by the G-22 and G-33 to reduce the world economy's susceptibility to crises showed the potential benefits of a regular international consultative forum embracing the emerging-market countries. Such a regular dialogue with a constant set of partners was institutionalized by the G-20 creation in 1999.

3. **How does the G-20 differ from the G-7?**

   The G-7 was established in 1976 as an informal forum of seven major industrial economies: Canada, France, Germany, Italy, Japan, the United Kingdom and the United States of America. The G-7 conducts dialogue and seeks agreement on current economic issues on the basis of the comparable interests of those countries. The G-20 was established in

1999 and reflects the diverse interests of the systemically significant industrial and emerging-market economies. (see: About the G-20). It has a high degree of representativeness and legitimacy on account of its geographical composition (members are drawn from all continents) and its large share of global population (two-thirds) and world GNP (around 90 per cent). The G-20's broad representation of countries at different stages of development gives its consensus outcomes greater impact than those of the G-7.

4. **Can all member countries exert equal influence?**

Achieving consensus is the underlying principle of G-20 activity with regard to comments, recommendations and measures to be adopted. There are no formal votes or resolutions on the basis of fixed voting shares or economic criteria. Every G-20 member has one 'voice' with which it can take an active part in G-20 activity. To this extent the influence a country can exert is shaped decisively by its commitment.

5. **What are the criteria for G-20 membership?**

In a forum such as the G-20, it is particularly important for the number of countries involved to be restricted and fixed to ensure the effectiveness and continuity of its activity. There are no formal criteria for G-20 membership and the composition of the group has remained unchanged since it was established. In view of the objectives of the G-20, it was considered important that countries and regions of systemic significance for the international financial system be included. Aspects such as geographical balance and population representation also played a major part.

6. **How is the G-20 taking forward work remitted to Finance Ministers by Leaders?**

The G-20 Finance Ministers were tasked from the Pittsburg Summit to take forward work in the following areas:

- Framework for Strong, Sustainable, and Balanced Growth
- Strengthening the International Financial Regulatory System
- Modernizing our Global Institutions to Reflect Today's Global Economy
- Reforming the Mandate, Mission, and Governance of the IMF
- Reforming the Mission, Mandate, and Governance of Our Development Banks
- Energy Security and Climate Change
- Strengthening Support for the Most Vulnerable

- Putting Quality Jobs at the Heart of the Recovery
- An Open Global Economy

CURRENT DEVELOPMENT IN G-20

The G-20 was established in 1999, in the wake of the 1997 Asian Financial Crisis, to bring together major advanced and emerging economies to stabilize the global financial market. Since its inception, the G-20 has held annual Finance Ministers and Central Bank Governors' Meetings and discussed measures to promote the financial stability of the world and to achieve a sustainable economic growth and development.

To tackle the financial and economic crisis that spread across the globe in 2008, the G-20 members were called upon to further strengthen international cooperation. Accordingly, the G-20 Summits have been held in Washington in 2008, in London and Pittsburgh in 2009, and in Toronto and Seoul in 2010.

The concerted and decisive actions of the G-20, with its balanced membership of developed and developing countries helped the world deal effectively with the financial and economic crisis, and the G20 has already delivered a number of significant and concrete outcomes:

First, the scope of financial regulation has been largely broadened and prudential regulation and supervision have been strengthened. There was also great progress in policy coordination thanks to the creation of the framework for a strong, sustainable and balanced growth designed to enhance macroeconomic cooperation among the G-20 members and therefore to mitigate the impact of the crisis. Finally, global governance has dramatically improved to better take into consideration the role and the needs of emerging of developing countries, especially through the ambitious reforms of the governance of the IMF and the World Bank.

Building on these important progresses, the G-20 has now to adapt to a new economic environment. It must prove that it is able to coordinate the economic policies of major economies on an ongoing basis.

2011 will be the occasion to build on the recent successes of the G-20 and ensure an active follow-up on processes already underway. It will also be the time to address other essential issues which are crucial to global stability such as the reform of the international monetary system and the volatility of commodity prices.

SOUTH ASIAN ASSOCIATION FOR REGIONAL COOPERATION (SAARC COUNTRIES)

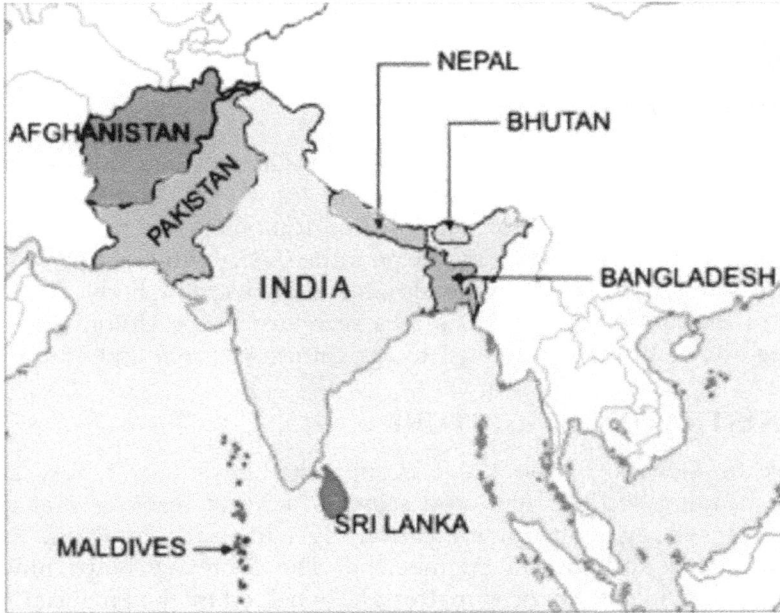

Established on December 8, 1985, South Asian Association for Regional Cooperation, popularly known as SAARC. SAARC is a unique concept. This esteem group is the one combined vision of eight different countries of Southeast Asia like Afghanistan, Bangladesh, Bhutan, India, Maldives, Nepal, Pakistan and Sri Lanka. These SAARC nations are home to nearly 1.5 billion people or about 22% of world's population.

So, the Heads of the State or Government of these aforementioned countries has created SAARC, keeping in view the welfare of the people, peace, stability and progress of South Asian region by fostering mutual understanding, meaningful cooperation and good neighbourly relations among these nations.

AGRICULTURE IN SAARC COUNTRIES

ICAR Play important role of India in SAARC

Indian Council of Agricultural Research or the ICAR is the autonomous and supreme body in the field of agriculture. Area of its work includes research and education related to agriculture. Through its vast network of thousands of employess, the ICAR has been working for the promotion of agricultural research in India.

GENERAL BODY

The General Body, presided by the Union Agriculture Minister, is the supreme authority of the ICAR. Other members of the General Body include the Minister of Animal Husbandry and Fisheries, senior provincial government officials, members of parliament, various scientific organizations and farmers.

GOVERNING BODY

The Governing Body, led by the Director-General, functions as the chief executive and decision making authority of the ICAR. Its members include agricultural scientists, legislators, eminent educationists and representatives of farmer community. Standing Finance Committee, Accreditation Board, Regional Committees and many scientific panels assist the governing body. The Director-General of the ICAR also functions as a secretary to the Union Government, providing advice in matters related to agriculture research and education.

ADMINISTRATIVE STRUCTURE

Director-General of the ICAR is supported by eight Deputy Director-Generals heading sections like crop sciences, natural resource management, animal sciences, agricultural education, agricultural extension, fisheries, horticulture and agricultural engineering. The Secretary helps him in the administration while in financial matters; he is helped by the Financial Advisor.

VAST NETWORK

The ICAR, employing about 30,000 people, has its offices and research centers in all parts of India. Several institutes, bureaus and project directorates work under the purview of ICAR. Thousands of agricultural scientists work in projects of the ICAR. The agricultural universities are also connected to the ICAR and take active part in teaching, research and education.

RECRUITMENT

Agricultural Scientists' Recruitment Board (ASRB) works as an autonomous recruiting agency of the ICAR for its Agricultural Research Services, equivalent technical and research management posts. National Academy of Agricultural Research Management (NAARM) provides required training to new recruits.

MGNREGA

The Mahatma Gandhi National Rural Employment Guarantee Act (MGNREGA) aims at enhancing the livelihood security of people in rural areas by guaranteeing hundred days of wage-employment in a financial year to a rural household whose adult members volunteer to do unskilled manual work.

The **Mahatma Gandhi National Rural Employment Guarantee Act** (MNREGA) is an Indian job guarantee scheme, enacted by legislation on August 25, 2005. The scheme provides a legal guarantee for one hundred days of employment in every financial year to adult members of any rural household willing to do public work-related unskilled manual work at the statutory minimum wage of Rs. 100 (US$ 2.17) per day in 2009 prices. The Central government outlay for scheme is Rs. 40,000 crore (US$ 8.68 billion) in FY 2010-11.

This act was introduced with an aim of improving the purchasing power of the rural people, primarily semi or unskilled work to people living in rural India, whether or not they are below the poverty line. Around one-third of the stipulated work force is women. The law was initially called the **National Rural Employment Guarantee Act** (NREGA) but was renamed on 2 October 2009.

POLITICAL BACKGROUND

This act was brought about by the UPA coalition government supported by the left parties. The promise of this project is considered by many to be one of the major reasons for the re-election of the UPA in the Indian general election, 2009.

Dr. Jean Drèze, a Belgian born economist, at the Delhi School of Economics, has been a major influence on this project. A variety of people's movements and organizations actively campaigned for this act.

THE PLAN

The act directs state governments to implement MNREGA "schemes". Under the MGNREGA the Central Government meets the cost towards the payment of wage, 3/4 of material cost and some percentage of administrative cost. State Governments meet the cost of unemployment allowance, 1/4 of material cost and administrative cost of State council. Since the State Governments pay the unemployment allowance, they are heavily incentivized to offer employment to workers.

However, it is up to the State Government to decide the amount of unemployment allowance, subject to the stipulation that it not be less than 1/4th the minimum wage for the first 30 days, and not less than 1/2 the minimum wage thereafter. 100 days of employment (or unemployment allowance) per household must be provided to able and willing workers every financial year.

PROVISIONS UNDER NREGA

- Adult members of a rural household, willing to do unskilled manual work, may apply for registration in writing or orally to the local Gram Panchayat

- The Gram Panchayat after due verification will issue a Job Card. The Job Card will bear the photograph of all adult members of the household willing to work under NREGA and is free of cost.

- The Job Card should be issued within 15 days of application.

- A Job Cardholder may submit a written application for employment to the Gram Panchayat, stating the time and duration for which work is sought. The minimum days of employment have to be at least fourteen.

- The Gram Panchayat will issue a dated receipt of the written application for employment, against which the guarantee of providing employment within 15 days operates

- Employment will be given within 15 days of application for work, if it is not then daily unemployment allowance as per the Act, has to be paid liability of payment of unemployment allowance is of the States.

- Work should ordinarily be provided within 5 km radius of the village. In case work is provided beyond 5 km, extra wages of 10% are payable to meet additional transportation and living expenses

- Wages are to be paid according to the Minimum Wages Act 1948 for agricultural labourers in the State, unless the Centre notices a wage rate which will not be less than Rs. 60 (US$ 1.3) per day. Equal wages will be provided to both men and women.

  Note : The original version of the Act was passed with Rs 60/ day as the minimum wage that needs to be paid under NREGA. However, a lot of states in India already have wage regulations with minimum wages set at more than Rs. 100 (US$ 2.17) per day. NREGA's minimum wage has since been changed to Rs. 100 (US$ 2.17) per day.

- Wages are to be paid according to piece rate or daily rate. Disbursement of wages has to be done on weekly basis and not beyond a fortnight in any case

- At least one-third beneficiaries shall be women who have registered and requested work under the scheme.

- Work site facilities such as crèche, drinking water, shade have to be provided

- The shelf of projects for a village will be recommended by the gram Sabah and approved by the zilla panchayat.

- At least 50% of works will be allotted to Gram Panchayats for execution.

- Permissible works predominantly include water and soil conservation, afforestation and land development works.

- A 60:40 wage and material ratio has to be maintained. No contractors and machinery is allowed.

- The Central Government bears the 100 percent wage cost of unskilled manual labour and 75 percent of the material cost including the wages of skilled and semi skilled workers.

- Social Audit has to be done by the Gram Sabha.

- Grievance redressed mechanisms have to be put in place for ensuring a responsive implementation process.

- All accounts and records relating to the Scheme should be available for public scrutiny.

HISTORY

MNREGA was launched on February 2, 2006 from Anantapur in Andhra Pradesh. The project was implemented in phased manner covering 130 districts by year 2007-08.With its spread over 625 districts across the country, the premier flagship program of the UPA Government has raised the productivity, increased the purchasing power, reduced distress migration and helped in creation of durable assets in rural India. This project has a formidable impact on rural India by providing employment to 41 million households in year 2010-11. Also, It has strengthen the social and gender equality dimensions as 23% workers under the scheme are Scheduled Castes, 17% Scheduled Tribes and 50% women.

FUNDING

MNREGA started with an initial outlay of $2.5 bn (Rs 11300 cr) in year 2006-07.The funding has considerably been increased as shown in the table below:

Year	Total Outlay(TO)	Wage Expenditure (Percent of TO)
2006-07	$2.5 bn	66
2007-08	$2.6 bn	68
2008-09	$6.6 bn	67
2009-10	$8.68 bn	70
2010-11	$8.91 bn	71

IMPLEMENTATION

The Comptroller and Auditor General (CAG) of India, in its performance audit of the implementation of MGNREGA have found "significant deficiencies" in the implementation of the act. The plan was launched in February 2006 in 200 districts and eventually extended to cover 593 districts. 44,940,870 rural households were provided jobs under NREGA during 2008-09, with a national average of 48 working days per household. In recent times, nrega workers have faced problems due to delays in payment of wages, some of which have been pending for months.

EMPLOYMENT UNDER NREGAS IN 2010

Indian Minister of State for Rural Development Pradeep Jain said in a written reply to a question in Rajya Sabha on Tuesday that as of 30 June, a total of 179,43,189 families in the country have been provided employment under MGNREGS.

WORKS/ACTIVITIES

The MGNREGA achieves twin objectives of rural development and employment. The MGNREGA stipulates that works must be targeted towards a set of specific rural development activities such as water conservation and harvesting, afforestation, rural connectivity, flood control and protection much as construction and repair of embankments, etc. Digging of new tanks/ponds, percolation tanks and construction of small check dams are also given importance. The employed are given work such as land leveling, tree plantation, etc. First a proposal is given by the Panchayat to the Block Office and then the Block Office decides whether the work should be sanctioned.

FIVE YEAR PLANS OF INDIA

The economy of India is based in part on planning through its five-year plans, developed, executed and monitored by the Planning Commission. With the Prime Minister as the ex-officio Chairman, the commission has a nominated Deputy Chairman, who has rank of a Cabinet minister. Montek Singh Ahluwalia is currently the Deputy Chairman of the Commission. The tenth plan completed its term in March 2007 and the eleventh plan is currently underway. Prior to the Fourth plan, the allocation of state resources was based on schematic patterns rather than a transparent and objective mechanism, which lead to the adoption of the Gadgil formula in 1969. Revised versions of the formula have been used since then to determine the allocation of central assistance for state plans.

1. First Five-Year Plan, 1951–56

2. Second Five-Year Plan, 1956–61

3. Third Five-Year Plan, 1961–66

4. Fourth Five-Year Plan, 1969–74

5. Fifth Five-Year Plan, 1974–79

6. Sixth Five-Year Plan, 1980–85

7. Seventh Five-Year Plan, 1985–90

8. Period between 1989–91

9. Eighth Five-Year Plan, 1992–97

10. Ninth Five Year Plan, 1997–2002

11. Tenth Five-Year Plan, 2002–07

12. Eleventh Five-Year Plan, 2007–12

13. Twelfth Five-Year Plan, 2012-17

FIRST FIVE-YEAR PLAN, 1951–1956

The first Indian Prime Minister, Jawaharlal Nehru presented the first five-year plan to the Parliament of India on 8 December 1951. The plan addressed, mainly, the agrarian sector, including investments in dams and irrigation. The agricultural sector was hit hardest by the partition of India and needed urgent attention. The total planned budget of Rs. 206.8 billion (US$ 23.6 billion in the 1950 exchange rate) was allocated to seven broad areas: irrigation and energy (27.2 percent), agriculture and community development (17.4 percent), transport and communications (24 percent), industry (8.4 percent), social services (16.64 percent), land rehabilitation (4.1 percent), and for other sectors and services (2.5 percent). The most important feature of this phase was active role of state in all economic sectors. Such a role was justified at that time because immediately after independence, India was facing basic problems like- deficiency of capital and low capacity to save.

The target growth rate was 2.1 percent annual gross domestic product (GDP) growth; the achieved growth rate was 3.6 percent. During the first five-year plan the net domestic product went up by 15 percent. The monsoon was good and there were relatively high crop yields, boosting exchange reserves and the per capita income, which increased by 8 percent. National income increased more than the per capita income due to rapid population growth. Many irrigation projects were initiated during this period, including the Bhakra Dam and Hirakud Dam. The World Health Organization, with the Indian government, addressed children's health and reduced infant mortality, indirectly contributing to population growth.

At the end of the plan period in 1956, five Indian Institutes of Technology (IITs) were started as major technical institutions. University Grant Commission was set up to take care of funding and take measures to strengthen the higher education in the country. Contracts were signed to start five steel plants; however these plants did not come into existence until the middle of the second five-year plan.

SECOND FIVE-YEAR PLAN, 1956–61

This plan functioned on the basis of a nude model. The Mahalanobis model was propounded by Prasanta Chandra Mahalanobis in the year 1953. The second five-year plan focused on industry, especially heavy industry. Unlike the First plan, which focused mainly on agriculture, domestic production of industrial products was encouraged in the Second plan, particularly in the development of the public sector. The plan followed the Mahalanobis model, an economic development model developed by the Indian statistician Prasanta Chandra Mahalanobis in 1953. The plan attempted to determine the optimal allocation of investment between productive sectors in order to maximize long-run economic growth. It used the prevalent state of art techniques of operations research and optimization as well as the novel applications of statistical models developed at the Indian Statiatical Institute. The plan assumed a closed economy in which the main trading activity would be centered on importing capital goods.

Hydroelectric power projects and five steel mills at Bhilai, Durgapur, and Rourkela were established. Coal production was increased. More railway lines were added in the north east.

The Atomic Energy Commission was formed in 1958 with Homi J. Bhabha as the first chairman. The Tata Institute of Fundamental Research was established as a research institute. In 1957 a talent search and scholarship program was begun to find talented young students to train for work in nuclear power.

The total amount allocated under the second five year plan in India was Rs. 4,800 crore. This amount was allocated among various sectors:

- Mining and industry
- Community and agriculture development
- Power and irrigation
- Social services
- Communications and transport
- Miscellaneous
- Third Five-Year Plan, 1961–66

The third plan stressed on agriculture and improving production of rice, but the brief Sino-Indian War of 1962 exposed weaknesses in the economy and shifted the focus towards the Defence industry. In 1965-66, India fought a war with Pakistan. The war led to inflation and the priority was shifted to price stabilization. The construction of dams continued. Many cement and fertilizer plants were also built. Punjab began producing an abundance of wheat.

Many primary schools were started in rural areas. In an effort to bring democracy to the grass root level, Panchayat elections were started and the states were given more development responsibilities.

State electricity boards and state secondary education boards were formed. States were made responsible for secondary and higher education. State road transportation corporations were formed and local road building became a state responsibility. The target growth rate of GDP (gross domestic product) was 4.5 percent. The achieved growth rate was 4.3 percent.

FOURTH FIVE-YEAR PLAN, 1969–74

At this time Indira Gandhi was the Prime Minister. The Indira Gandhi government nationalized 14 major Indian banks and the Green Revolution in India advanced agriculture. In addition, the situation in East Pakistan (now Bangladesh) was becoming dire as the Indo-Pakistani War of 1971 and Bangladesh Liberation War took place.

Funds earmarked for the industrial development had to be diverted for the war effort. India also performed the Smiling Buddha underground nuclear test in 1974, partially in response to the United States deployment of the Seventh Fleet in the Bay of Bengal. The fleet had been deployed to warn India against attacking West Pakistan and extending the war.

FIFTH FIVE-YEAR PLAN, 1974–79

Stress was laid on employment, poverty alleviation, and justice. The plan also focused on self-reliance in agricultural production and defence. In 1978 the newly elected Morarji Desai government rejected the plan. Electricity Supply Act was enacted in 1975, which enabled the Central Government to enter into power generation and transmission leaders.

The Indian national highway system was introduced for the first time and many roads were widened to accommodate the increasing traffic. Tourism also expanded.

SIXTH FIVE-YEAR PLAN, 1980–85

The sixth plan also marked the beginning of economic liberalization. Price controls were eliminated and ration shops were closed. This led to an increase in food prices and an increase in the cost of living. This was the end of Nehruvian Plan and Rajiv Gandhi was prime minister during this period.

Family planning was also expanded in order to prevent overpopulation. In contrast to China's strict and binding one-child policy, Indian policy did not rely on the threat of force. More prosperous areas of India adopted family planning more rapidly than less prosperous areas, which continued to have a high birth rate.

SEVENTH FIVE-YEAR PLAN, 1985–90

The Seventh Plan marked the comeback of the Congress Party to power. The plan laid stress on improving the productivity level of industries by upgrading of technology.

The main objectives of the 7th five year plans were to establish growth in areas of increasing economic productivity, production of food grains, and generating employment opportunities.

As an outcome of the sixth five year plan, there had been steady growth in agriculture, control on rate of Inflation, and favourable balance of payments which had provided a strong base for the seventh five Year plan to build on the need for further economic growth. The 7th Plan had strived towards socialism and energy production at large. The thrust areas of the 7th Five year plan have been enlisted below:

- Social Justice
- Removal of oppression of the weak
- Using modern technology
- Agricultural development
- Anti-poverty programs
- Full supply of food, clothing, and shelter
- Increasing productivity of small and large scale farmers
- Making India an Independent Economy

Based on a 15-year period of striving towards steady growth, the 7th Plan was focused on achieving the pre-requisites of self-sustaining growth by the year 2000. The Plan expected a growth in labour force of 39 million people and employment was expected to grow at the rate of 4 percent per year.

Some of the expected outcomes of the Seventh Five Year Plan India are given below:

- Balance of Payments (estimates): Export - Rs. 33,000 crore (US$ 7.2 billion), Imports - (-) 54,000 crore (US$ 11.7 billion), Trade Balance - (-) 21,000 crore (US$ 4.6 billion)
- Merchandise exports (estimates): 60,653 crore (US$ 13.2 billion)
- Merchandise imports (estimates): 95,437 crore (US$ 20.7 billion)
- Projections for Balance of Payments: Export - 60,700 crore (US$ 13.2 billion), Imports - (-) 95,400 crore (US$ 20.7 billion), Trade Balance- (-) 34,700 crore (US$ 7.5 billion)

Seventh Five Year Plan India strove to bring about a self-sustained economy in the country with valuable contributions from voluntary agencies and the general populace.

PERIOD BETWEEN 1989–91

1989-91 was a period of political instability in India and hence no five year plan was implemented. Between 1990 and 1992, there were only Annual Plans. In 1991, India faced a crisis in Foreign Exchange (Forex) reserves, left with reserves of only about US$ 1 billion. Thus, under pressure, the country took the risk of reforming the socialist economy. P.V. Narasimha Rao) was the twelfth Prime Minister of the Republic of India and head of Congress Party, and led one of the most important administrations in India's modern history overseeing a major economic transformation and several incidents affecting national security. At that time Dr. Manmohan Singh (currently, Prime Minister of India) launched India's free market reforms that brought the nearly bankrupt nation back from the edge. It was the beginning of privatisation and liberalisation in India.

EIGHTH FIVE-YEAR PLAN, 1992–97

Modernization of industries was a major highlight of the Eighth Plan. Under this plan, the gradual opening of the Indian economy was undertaken to correct the burgeoning deficit and foreign debt. Meanwhile India became a member of the World Trade Organization on 1 January 1995.This plan can be termed as Rao and Manmohan model of Economic development. The major objectives included, containing population growth, poverty reduction, employment generation, strengthening the infrastructure, Institutional building, tourism management, Human Resource development, Involvement of Panchayat raj, Nagarapalikas, N.G.O'S and Decentralization and people's participation. Energy was given priority with 26.6% of the outlay. An average annual growth rate of 6.7% against the target 5.6% was achieved.

NINTH FIVE YEAR PLAN, 1997–2002

Ninth Five Year Plan India runs through the period from 1997 to 2002 with the main aim of attaining objectives like speedy industrialization, human development, full-scale employment, poverty reduction, and self-reliance on domestic resources.

Background of Ninth Five Year Plan India: Ninth Five Year Plan was formulated amidst the backdrop of India's Golden Jubilee of Independence.

The main objectives of the Ninth Five Year Plan of India are:

- to prioritize agricultural sector and emphasize on the rural development
- to generate adequate employment opportunities and promote poverty reduction
- to stabilize the prices in order to accelerate the growth rate of the economy
- to ensure food and nutritional security
- to provide for the basic infrastructural facilities like education for all, safe drinking water, primary health care, transport, energy
- to check the growing population increase

- to encourage social issues like women empowerment, conservation of certain benefits for the Special Groups of the society

- to create a liberal market for increase in private investments

During the Ninth Plan period, the growth rate was 5.35 per cent, a percentage point lower than the target GDP growth of 6.5 per cent.

TENTH FIVE-YEAR PLAN, 2002–07

- Attain 8% GDP growth per year.

- Reduction of poverty ratio by 5 percentage points by 2007.

- Providing gainful and high-quality employment at least to the addition to the labour force.

- All children in India in school by 2003; all children to complete 5 years of schooling by 2007.

- Reduction in gender gaps in literacy and wage rates by at least 50% by 2007.

- Reduction in the decadal rate of population growth between 2001 and 2011 to 16.2%.

- Increase in Literacy Rates to 75 per cent within the Tenth Plan period (2002 to 2007).

ELEVENTH FIVE-YEAR PLAN, 2007–12

The eleventh plan has the following objectives:

1. **Income and Poverty**

   - Accelerate GDP growth from 8% to 10% and then maintain at 10% in the 12th Plan in order to double per capita income by 2016-17

   - Increase agricultural GDP growth rate to 4% per year to ensure a broader spread of benefits

   - Create 70 million new work opportunities.

   - Reduce educated unemployment to below 5%.

   - Raise real wage rate of unskilled workers by 20 percent.

   - Reduce the headcount ratio of consumption poverty by 10 percentage points.

2. **Education**

   - Reduce dropout rates of children from elementary school from 52.2% in 2003-04 to 20% by 2011-12

- Develop minimum standards of educational attainment in elementary school, and by regular testing monitor effectiveness of education to ensure quality
- Increase literacy rate for persons of age 7 years or above to 85%
- Lower gender gap in literacy to 10 percentage point
- Increase the percentage of each cohort going to higher education from the present 10% to 15% by the end of the plan

3. **Health**
- Reduce infant mortality rate to 28 and maternal mortality ratio to 1 per 1000 live births
- Reduce Total Fertility Rate to 2.1
- Provide clean drinking water for all by 2009 and ensure that there are no slip-backs
- Reduce malnutrition among children of age group 0-3 to half its present level
- Reduce anaemia among women and girls by 50% by the end of the plan

4. **Women and Children**
- Raise the sex ratio for age group 0-6 to 935 by 2011-12 and to 950 by 2016-17
- Ensure that at least 33 percent of the direct and indirect beneficiaries of all government schemes are women and girl children
- Ensure that all children enjoy a safe childhood, without any compulsion to work

5. **Infrastructure**
- Ensure electricity connection to all villages and BPL households by 2009 and round-the-clock power.
- Ensure all-weather road connection to all habitation with population 1000 and above (500 in hilly and tribal areas) by 2009, and ensure coverage of all significant habitation by 2015
- Connect every village by telephone by November 2007 and provide broadband connectivity to all villages by 2012
- Provide homestead sites to all by 2012 and step up the pace of house construction for rural poor to cover all the poor by 2016-17

6. **Environment**
- Increase forest and tree cover by 5 percentage points.
- Attain WHO standards of air quality in all major cities by 2011-12.
- Treat all urban waste water by 2011-12 to clean river waters.
- Increase energy efficiency by 20 percentage points by 2016-17.

TWELFTH FIVE-YEAR PLAN, 2012–17

1. **Monitorable Targets:** e.g. Average GDP Growth of 8 percent. • Agriculture Growth of 4 percent. Reducing head-count poverty by 10 percentage point. Generating 50 million work opportunities. • Eliminating gender social gap in education. • Reducing IMR to 25, MMR to 100 and TFR to 2.1. Enhance infrastructure investment to 9% of GDP. • Achieve universal road connectivity and access to power for all villages. • Access to banking services for 90 percent households. • Major welfare benefits and subsidies via Aadhaar.

2. **Strategy for Achieving Inclusiveness:** •There are two routes to inclusiveness:- (*a*) through higher growth which expands income and employment opportunities (*b*) through government pro-poor programmes which target poorer groups. Both are important. • Twelfth Plan combines the pro-poor programme approach with efforts to get a growth pattern which is faster and inherently more inclusive. The two routes mutually reinforcing. High growth generates more revenues to finance inclusiveness programmes. • Many inclusiveness programmes (health and education) contribute to growth.

3. **Macro-Economic Challenges in Achieving High Growth:** • Must increase the rate of investment especially in infrastructure. • Domestic savings must increase even more to reduce the investment saving gap which is necessary to keep the Current Account Deficit under control. • Government dis-savings must be eliminated. This means fiscal deficit must be reduced. Growth of subsidies has to be contained. 75% of the investment in the economy is private investment (household plus corporate). • Both the Centre and the States have to create an eco-system that encourages private investment. • Infrastructure especially quality of power and availability of skills is critical.

4. **Effectiveness of Plan Programmes:** • Twelfth Plan sets ambitious targets for Flagship Programmes in areas of Health, Education, Rural Infrastructures, Livelihood Development etc. • Too much focus on the level of expenditure in these programmes. Not enough on effectiveness in terms of end results. Implementation in the field is the responsibility of State Government agencies. However, programme guidelines are set by the Central Government. There are demands for greater flexibility from states. • We are responding as follows: Guidelines are being made more flexible to meet the requirement of individual States. 10% flexi-fund within each scheme for innovations.

5. **Energy:** • 8 percent GDP growth requires 6 percent growth in energy supply from all sources. • Our fossil fuel resources are limited and our important dependence is already high. Enhanced Energy Efficiency in

all sectors. We need to push push for renewable Energy: Wind, Solar and Storage Hydro. • Difficult Policy Issues: Coal vs. Forest Cover, Petroleum Price Distortions, Natural Gas Exploration Framework. • All Energy Prices: Coal, Petroleum product, Natural Gas and Electric power are currently under priced. Fuel adjustment is needed.

6. **Water:** • Management of water resource is a major challenge. Estimates of water availability have been optimistic. • Twelfth Plan proposes modified AIBP and expanded Watershed Management Programme. • Water sector needs better Regulatory Framework. New Groundwater Law. Water Regulatory Authorities in each state. Natural Water Framework Law. Agriculture accounts for 80% of water use at present, must shift to more to more water efficient agriculture practices. • Manage urban and industrial water demand through recycling and rationalise user charges.

7. **Alternative Scenarios:** • 12th Plan goal of 8% inclusive growth is not a foregone conclusion. Depends on difficult policy decisions to be taken by Centre and States. • For the first time Plan presents three scenarios Strong Inclusive Growth 8% Insufficient Action 6 to 6.5% Policy logjam 5 to 5.5%. • Anything much less than 8% will not satisfy aspirations of the people.

REFERENCES

Adhikary, M.M., Acharya, S.K. and Basu, D. 2011. **Extension Strategy on Natural Resource Management.** NIAB. P. 624.

Bhargav, Kapil & Nidhi Jethwant (2010). **Cooperation Management and Rural Development.** Daya Publishing House, New Delhi.

Dubey, Virendra Kumar and K Chaturvedi (2010). **Relevance of Ancient Indian Agricultural Wisdom in Modern Context.** Daya Publishing House, New Delhi.

Gupta, Om and Sudesh Sharma (2010). **Media and Communication Research: Changing Paradigms.** Daya Publishing House, New Delhi.

Hansra, B.S.: ed. (2003). **Agribusiness and Extension Management.** New India Publishing Agency New Delhi.

Jain, S.C. (2010). **New Trends in Personnel Management.** Daya Publishing House, New Delhi.

Ngurtinkhuma, R.K. (2010). **Public Library in India: Impact on Socio-Cultural and Educational Development of Mizoram.** Daya Publishing House, New Delhi.

Patil, D.A., Dhrere, A.M., and Pawar, C.B. 2010. **Communication and development in India.** Rajat Publications New Delhi. P. 382.

Parikh, J.C. (2010). **Educational Psychology.** Daya Publishing House, New Delhi.

Rahim, Mohammad & M.S. Nataraju (2010). **Agricultural Education: The Occupational Aspirations.** Daya Publishing House, New Delhi.

Rahman, H. (2010). **Entrepreneurship in Agriculture: Scopes and Opportunities.** Daya Publishing House, New Delhi.

Rathakrishnan, T., Thomas, M.I. and Nirmala, L. 2009. **Advances and Challenges in Agricultural Extension and Rural Development.** New India Publishing Agency, Pitampura, New Delhi. P. 452.

Rathakrishnan, R.T. (2009). **Advances and Challenges in Agricultural Extension and Rural Development.** New India Publishing Agency New Delhi.

Ray, G.L. 2011. **Extension Communication and Management.** Kalyani Publishers, New Delhi. Pp. 358.

Sannigrahi, Asoke (2011). **Agriculture and Waste Management for Sustainable Future.** New India Publishing Agency New Delhi.

Santha Govind, Tamilselvi, G. and Meenambigai, J. 2011. **Extension Education and Rural Development.** Agrobios Jodhpur. P. 493.

Shetty, Karun (2010). **Changing Face of Journalism in Digital Age.** Daya Publishing House, New Delhi.

Shetty, Karun (2010). **Communication for Social Change.** Daya Publishing House, New Delhi.

Singh, M.K. (2010). **Fundamental of Media.** Daya Publishing House, New Delhi.

Tripathy, Manoranjan (2010). **Public Relations: Bridging Technologies and Monitoring Public and the Media.** Daya Publishing House, New Delhi.